MATHEMATICS RESEARCH DEVELOPMENTS

SEVERAL PROBLEMS OF APPLIED MATHEMATICS AND MECHANICS

MATHEMATICS RESEARCH DEVELOPMENTS

Additional books in this series can be found on Nova's website
under the Series tab.

Additional e-books in this series can be found on Nova's website
under the e-book tab.

MATHEMATICS RESEARCH DEVELOPMENTS

SEVERAL PROBLEMS OF APPLIED MATHEMATICS AND MECHANICS

IVANE GORGIDZE
AND
TAMAR LOMINADZE
EDITORS

nova publishers
New York

For permission to use material from this book please contact us:
Telephone 631-231-7269; Fax 631-231-8175
Web Site: http://www.novapublishers.com

LIBRARY OF CONGRESS CATALOGING-IN-PUBLICATION DATA

Several problems of applied mathematics and mechanics / [edited by] Ivane Gorgidze and Tamar Lominadze.
p. cm.
Includes bibliographical references and index.
ISBN: 978-1-62081-603-5 (hardcover)
1. Mechanics, Applied. I. Gorgidze, Ivane. II. Lominadze, Tamar.
TA350.S48 2011
620.1--dc23

2012013587

Published by Nova Science Publishers, Inc. † New York

Dedicated to the 105th Birth Anniversary of Professor Alexi Gorgidze

CONTENTS

PREFACE

This book aims to present the newest research in the fields of mathematics and mechanics done by outstanding Georgian scientists. mathematics and mechanics are fundamental sciences. Theories of mathematics and mechanics are the basis of modern technological improvements and, therefore, interest towards them are increasingly important. This book presents the latest works of outstanding Georgian scientist and mathematician, academician Nikoloz Muskhelishvili. Together with the works of N. Muskhelishvili, his book includes the recent works of a number of modern Georgian scientists who work in the fields of mathematics and mechanics.

The first part of Chapter 1 investigates some new properties associated with the mathematical models of v*on Kármán-Mindlin-Reissner (KMR)*-type systems of partial differential equations (PDEs). For completeness, the authors also lead 2D *Reissner-Filon*-type models for elastic plates of variable thicknesses.

The second part of the chapter studies approximate methods for solving the corresponding boundary value problems (BVPs). The authors consider the continuous analogue of Peaceman-Rechford's Alternating Direction Method (ADM) and the high-order accuracy schemes for approximate solutions according to the finite-element method by using a new class of B-type spline-functions. The exactness of such schemes depends on differential properties of unknown solutions: it has an arbitrary order of accuracy in respect to a mesh width in cases of sufficiently smooth functions and *Sard*-type best coefficients, characterizing the remainder proximate members on a less smooth class of admissible solutions. In this connection for Cauchy problem suggests new schemes having arbitrary order of accuracy and based on Gauss-Hermite processes. These processes are new, even for ordinary differential equations.

Important problem areas of fuel and energy complex based on the cognitive approach have been determined in Chapter 2. The system analysis, taking into consideration strong and weak properties, risks and chances, environmental, political, legal, social, economic, psychological and other factors, has been conducted. The methodology of constructing cognitive models is proposed. As a result of modeling, the possible scenarios of development of the energy economy have been investigated.

In Chapter 3, using spinor representation of orthogonal transformations, the expressions between second order complex unitary transformations matrixes and real orthogonal matrixes of spatial rotations in three dimensional Euclidean space L^3 are received, which allows easily calculating of corresponding Euler's angles. The obtained results have enabled reducing the actually three-dimensional problem of spatial motion control to the one-dimensional problem; control kinematical functions of Euler's angles and control spinor matrix of rotation were constructed, by means of which the control process of spatial rotations is completely determined.

The influence of the assumption about the existence of cardinal coefficients of criteria importance, consistent with the importance ordering of criteria (according to the definitions in the criteria importance theory), on the preference relation generated by this information, is investigated in Chapter 4.

In Chapter 5, the mixed problem with first order boundary conditions for systems of equations of parabolic types are considered:

$$B\frac{\partial u}{\partial t} = Lu + f,$$

where B - positively defined, symmetric, and bounded matrix, L - strong elliptic operator with variable coefficients, containing the mixed derivatives, $u = \left(u^{(1)}, u^{(2)}, \dots, u^{(n)}\right)$ and $f = \left(f^{(1)}, f^{(2)}, \dots, f^{(n)}\right)$ - n-dimensional vectors. The constructed three-layer factorized scheme's solution requires no inversion of matrix B. The received algorithms can be effectively used for multiprocessing computing systems. For difference schemes, the aprioristic estimation on layer in the norm of mesh space $\overset{0}{W}{}_2^{(1)}$ is received, on which the basis of the solution convergence of the difference scheme to the solution of an initial problem is proved.

In Chapter 6, a mathematical model has been elaborated to explore pneumatic equalizing spring suspension dynamics, which represents a system of differential equations and describes current processes of thermodynamical, thermal and mass transmission in the research of pneumatic systems.

The authors call an infinite sequence (ξ_k) of random variables given on a probability space (Ω, A, P) a Denjoy-Luzin system if for an infinite sequence (x_k) of scalars the almost sure (a.s.) convergence of the series $\sum_{k=1}^{\infty} |x_k \xi_k|$ implies $\sum_{k=1}^{\infty} |x_k| < \infty$. In Chapter 7, the authors give a characterization of Denjoy-Luzin systems. Moreover, the authors show that if (ξ_k) is a Denjoy-Luzin system and (x_k) is an infinite sequence of elements of a Banach space X then a.s. unconditional convergence of the series $\sum_{k=1}^{\infty} x_k \xi_k$ implies unconditional convergence of the series $\sum_{k=1}^{\infty} x_k$ in X. The authors show also that a similar result for systems of absolute convergence is true only for finite dimensional Banach spaces.

The Cauchy problem with closed support for quasi-linear hyperbolic equation with admissible parabolic degeneracy is investigated in Chapter 8. The structure of the domain of definition of the solution is described and sufficient conditions for existence of sub-domains impenetrable for that solution are founded.

In Chapter 9, a new result is obtained on the vanishing of local time for a non-negative continuous semimartingale at zero. Based on this result, an early exercise premium representation of a value function for the American put option is obtained in a one-dimensional general diffusion model.

General questions regarding optimal stopping for an inhomogeneous Markov process for time-dependent gain function are investigated. The connection between optimal stopping problems for an inhomogeneous standard Markov process and the corresponding homogeneous Markov process constructed in the extended state space is established. A detailed characterization of a value-function and the limit procedure for its construction in the problem of optimal stopping of an inhomogeneous Markov process is given in Chapter 10. The form of ε-optimal (optimal) stopping times is also found.

Chapter 11 focuses on the actual practical information regarding boundary problems with the boundary singularities. Dirichlet generalized boundary problem for a Laplace equation is considered for both finite and infinite domains. The case of a boundary function with a limited number of first kind break points is considered under this generalized boundary problem. In this chapter, one method is a given ordinary problem for the reduction of the Dirichlet generalized boundary problem with an harmonic function. The method is constructed on the basis of fictitious sources and is applied to finite and infinite domains. From the authors'point of view, this method is characterized by simplicity, stability, and allows for high accuracy. Also, it is oriented for a wide range of users (especially for the researchers in engineering problems). Examples with cusp points in the role of break points are also considered. The results of numerical experiments are given.

In Chapter 12, the authors show that in an arbitrary separable Banach space X there exists an universal series $\sum_i x_i$ with the following additional properties: $\sum_{i=1}^{\infty} \|x_i\|^p < \infty$ for each $p \in [1, \infty]$; the series $\sum_i \vartheta_i x_i$ converges in X for almost all sequences of signs (ϑ_i). In a wide class of complete separable metrizable topological vector spaces the existence of an universal series with property (2) is established as well.

In Chapter 13, a structural pre-sheaf of C^*- algebras, which is also a co-pre-sheaf of C^*-algebras is constructed on the space $\Pr imU$ of primitive ideals of a C^*- algebra U. Also, co-chain and chain complexes with coefficients in a pre-sheaf and a co-pre-sheaf, respectively, are considered. A sheaf on $\Pr imU$ and pre-sheaves on PU and $SpaseU$ are constructed from the obtained pre-sheaf by stanstand methods.

In Chapter 14 the new concept of mixed extensional lexicographic bimatrix game $\Gamma = (\Gamma^1, ..., \Gamma^r)$ of finite dimension is discussed. In this extension the players choose the components $\Gamma^1, ..., \Gamma^r$ of the game Γ with given probabilistic distributions on the set of such scalar game-criteria. The concept of the situation, a situation in the coordinated strategies, an equilibrium situation with positive weights of preferences is given in a lexicographic game. The conditions of the existence of such equilibrium situation are proved in a lexicographic bimatrix game.

In Chapter 15, a mathematical model of an economic system is presented in the form of second order neutral differential equation. Necessary conditions of optimality are established for the corresponding neutral optimal control problem.

Chapter 16 considers the approximation method of solution of unsteady boundary layer equation of free convection weakly conducting degree fluid.

By using the method of successive approximation, Chapter 17 studies the unsteady problem of an infinitely rotating porous plate in a conducting fluid with account of heat transfer with variable electroconductivity $\sigma = \sigma_0 \frac{T}{T_\infty}$, when there is an injection of the same flow with $v_w(t)$ speed into the plate.

To determine the thickness of the dynamic and thermal boundary layers, differential equations are obtained and exact solutions were found in special cases when the injection velocity varies according to different laws and between the thicknesses of a functional dependence of the form $\delta_T(t) = \gamma\delta(t)$.

All physical characteristics are calculated.

By method of separation of variables, the functional solution of a number of boundary and boundary-contact problems on thermo electric elasticity equilibrium of single and multi-layer bodies, limited with coordinate surfaces of spherical system of coordinates, is constructed in Chapter 18. The body is affected by stationary temperature and electric fields and surface perturbations. The elastic body is considered as transtropic (transversally isotropic) with surface isotropy r=const.

In Chapter 19, the author presents the original approach to synthesize the one-way function and the key exchange algorithm via an open channel. In this work, the reader will find the one-way function's structure, first published in [1]. Furthermore, this one-way function was studied and reviewed in several publications (see [2 - 4]). These studies show that it is possible to build large sets of n x n-sized commutative matrices over the Galois field GF(p) for fixed natural values of n. The set is built as a cyclic multiplicative group of cardinality $2^n - 1$ with a specific generator-matrix. These sets of matrices are used in the construction of the one-way function as well as in the key exchange algorithm. Section 2 of the article represents the principally new one-way function, and section 3 addresses the new key exchange algorithm via an open channel.

In: Several Problems of Applied Mathematics and Mechanics ISBN: 978-1-62081-603-5
Editors: Ivane Gorgidze and Tamar Lominadze

Chapter 1

SOME REMARKS ON RELATIVELY REFINED THEORIES FOR ELASTIC PLATES

Tamaz Vashakmadze
I. Vekua Institute of Appl. Mathematics of Iv. Javakhishvili
Tbilisi State University, Tbilisi, Georgia

Abstract

The first part of the chapter investigates some new properties associated with the mathematical models of *von Kármán-Mindlin-Reissner (KMR)*-type systems of partial differential equations (PDEs). For completeness, we also lead 2D *Reissner-Filon*-type models for elastic plates of variable thicknesses.

The second part of the chapter studies approximate methods for solving the corresponding boundary value problems (BVPs). We consider the continuous analogue of Peaceman-Rechford's Alternating Direction Method (ADM) and the high-order accuracy schemes for approximate solutions according to the finite-element method by using a new class of B-type spline-functions. The exactness of such schemes depends on differential properties of unknown solutions: it has an arbitrary order of accuracy in respect to a mesh width in cases of sufficiently smooth functions and *Sard*-type best coefficients, characterizing the remainder proximate members on a less smooth class of admissible solutions. In this connection for Cauchy problem suggests new schemes having arbitrary order of accuracy and based on Gauss-Hermite processes. These processes are new, even for ordinary differential equations.

Keywords: Elasticity, Plate, Physical soundness, Finite-difference scheme, Gauss quadrature formula, Mesh width, Alternating Direction Method

Dedication

This work is dedicated to the anniversary of my teacher Alex Gorgidze for considering some BVPs of two-dimensional theory of elasticity connected with the fundamental works of N. Muskhelishvili, I. Vekua, and A. Gorgidze [1-3].

1. Mathematical Models of *von Kármán*-Type Systems for DEs

One of the most principal objects in the development of mechanics and mathematics is a system of nonlinear differential equations for an elastic isotropic plate constructed by *von Kármán*. This system represents the most essential part of the main manuals in elasticity theory. In spite of this, in 1978, *Truesdell* expressed an idea about the neediness associated with the "Physical Soundness" of *von Kármán's* system. This circumstance generated the problem of justifying *von Kármán's* system. Consequently, this problem was studied by many authors, but Ciarlet investigated it with the most attention [4]. In particular, he wrote, "The *von Kármán* equations may be given a full justification by means of the leading term of a formal asymptotic expansion" [4, p.368]. This result is obviously not sufficient for a justification of the "Physical Soundness" of *von Kármán's* system as representations by asymptotic expansions are dissimilar: leading terms are only coefficients of power series without any physical meaning.

Based on the [5, 6], the method of constructing such an isotropic, inhomogeneous, 2D, nonlinear model of *a von Kármán- -Mindlin-Mindlin-Reissner (K MindlinMR)*type for elastic plates with variable thickness is given, the terms of which possess a quite determined "Physical Soundness". The corresponding variables are quantities with certain physical meanings: averaged components of the displacement vector, bending and twisting moments, shearing forces, rotation of normals, surface efforts. In addition, the corresponding equations are constructed taking into account the equality conditions of the main vector and moment to zero. By choosing parameters in the isotropic case from the *K MindlinMR Mindlin*-type system (having a continuum power), the *von Kármán* system is obtained as one of the possible models. The given method differs from the classical one by the fact that, according to the classical method, one of the equations of the *von Kármán* system represents one of *S.t-Venant's* compatibility conditions: it's obtained on the basis of geometry and does not take into account the equilibrium equations. This remark is essential for dynamical problems.

Using the methodology of [5] from chapter 1 (in the case when the thin-walled structure is an elastic isotropic homogeneous plate with constant thickness), we have the following nonlinear systems for PDEs of the *KMR* type:

$$D\Delta^2 \bar{u}_3 = \left(1-\frac{h^2(1+2\gamma)(2-\nu)}{3(1-\nu)}\Delta\right)(g_3^+ - g_3^-) + 2h\left(1-\frac{2h^2(1+2\gamma)}{3(1-\nu)}\Delta\right)[\bar{u}_3,\Phi^*] + h(g_{3,\alpha}^+ - g_{3,\alpha}^-)$$
$$-\int_{-h}^{h}\left(zf_{\alpha,\alpha} - (1-\frac{1}{1-\nu}\Delta(h^2-z^2)f_3)\right)dz + R_1[\bar{u}_3;\gamma] \quad (1)$$

$$\Delta^2\Phi^* = -\frac{E}{2}[\bar{u}_3,\bar{u}_3] + \frac{\nu}{2}\Delta(g_3^+ + g_3^-) + \frac{1+\nu}{2h}\bar{f}_\alpha + R_2[\Phi^*] \quad (2)$$

$$Q_{\alpha 3} - \frac{1+2\gamma}{3}h^2\Delta Q_{\alpha 3} = -D\Delta\bar{u}_{3,\alpha} + \frac{h^2(1+2\gamma)}{3(1-\nu)}\partial_\alpha\left(g_3^+ - g_3^- + 2h(1+\nu)\right)[\bar{u}_3,\Phi^*]$$
$$+h(g_\alpha^+ - g_\alpha^-) - \int_{-h}^{h} zf_\alpha dz + \frac{1+\nu}{2(1-\nu)}\int_{-h}^{h}(h^2-z^2)f_{3,\alpha}dz + R_{2+\alpha}[Q_{\alpha 3};\gamma]. \quad (3)$$

The system (1) - (3), without reminder terms R, gives a 2D system of refined theories with control parameters γ. By choosing γ, we got all well-known, refined theories and some new ones from other γ. The constructed models, together with certain independent scientific interests, represent such a form of spatial models, which allow not only to construct, but also to justify the von KMR-type systems in the stationary, as well as in the non-stationary cases. We are reminded that even in the case of an isotropic elastic plate with constant thickness, the subject of justification constituted an unsolved problem. The point is that *von Kármán, Love, Timoshenko, Landau & Lifshits,* etc. considered the compatibility conditions of *St.-Venant-Beltrami* as one of the equations of the corresponding system. This fact was verified also recently by *Podio-Guidugli.*

In the presented model, we demonstrated a correct equation that is especially important for dynamic problems. In this case, along the quantities describing the vertical directions and surface wave processes, it is necessary to take into account the quantity $\Delta\partial_{tt}\Phi$ as it corresponds to wave. This remark is essential for dynamical problems. Furthermore, in this case, along the quantities describing the vertical directions and surface wave processes, it is necessary to take into account the quantity $\Delta\partial_{tt}\Phi$ as it corresponds to wave processes in the horizontal direction. Moreover, in this case, along the quantities describing the vertical directions and surface wave processes it is necessary to take into account the quantity $\Delta\partial_{tt}\Phi$ as it corresponds to wave the equilibrium equations. This remark is also essential for dynamical problems. Furthermore, for isotropic and generalized transversal elastic plates in linear equations, this remark is essential for dynamical problems. The equations have the following form [6]:

$$
\begin{aligned}
&\left(D\Delta^2 + 2h\rho\partial_{tt} - 2DE^{-1}(1+\upsilon)\rho\partial_{tt}\Delta\right)w = \\
&\left(1 - \frac{h^2(1+2\gamma)(2-\upsilon)}{3(1-\upsilon)}\Delta\right)\left(g_3^{\,+} - g_3^{\,-}\right) \\
&+ 2h\left(1 - \frac{2h^2(1+2\gamma)}{3(1-\upsilon)}\Delta\right)\left[u_3^{\,*}, F_*\right] + h\left(g_{\alpha,\alpha}^{\,+} - g_{\alpha,\alpha}^{\,-}\right)
\end{aligned}
\tag{4}
$$

$$
\begin{aligned}
&\left(\Delta^2 - \frac{1-\nu^2}{E}\rho_1\Delta\partial_{tt}\right)\Phi = \\
&-\frac{E}{2}[w,w] + \frac{\nu}{2}\left(\Delta - \frac{2\rho_1}{E}\partial_{tt}\right)\left(g_3^{+} + g_3^{-}\right) + \frac{1+\nu}{2h}f_{\alpha,\alpha}
\end{aligned}
\tag{5}
$$

From (4)-(5) follows *von Kármán* equations if in (4) $\gamma = -0.5, g_\alpha^{\pm} = 0$ and in (5) $f_\alpha = \rho_1 = \Delta g_3^{\pm} = 0$. In addition, an equation corresponding to (5) by *von Kármán A. Föppl,Love, Lukasievicz, Tomoshenko, Donnel, Landau, Ciarlet, Antman*, etc. was constructed by the condition $\varepsilon_{11,22} - 2\varepsilon_{12,12} + \varepsilon_{22,11} = -0.5[u_3, u_3]$ and Hooke's law (but without using the equilibrium equations). As we prove in works [6, 7], the form type (5) follows immediately for more general cases, when thin-walled elastic structures are anisotropic and if we use Hooke's law, equilibrium equations with and nonlinear relations between the strain tensor and displacement vector: $\varepsilon_{\alpha\beta} = 0.5(u_{\alpha,\beta} + u_{\beta,\alpha} + u_{3,\alpha}u_{3,\beta})$. Now we shall prove that equation (5) has the following form in the dynamical case:

$$\left(-\frac{1-\nu^2}{E}\rho_1\Delta\partial_{tt}\right)\Phi = \frac{\nu}{2}\left(\Delta - \frac{2\rho_1}{E}\partial_{tt}\right)(g_3^+ + g_3^-) + \frac{1+\nu}{2h}f_{\alpha,\alpha}. \quad (6)$$

Thus, we must demonstrate that both ways are the same for the formed expression $\Delta^2\Phi - 0.5E[w,w]$.

In fact, when we constructed (5), it is true [6]:

$$(\lambda^* + 2\mu)\Delta(\bar{\varepsilon}_{11} + \bar{\varepsilon}_{22}) = (2\mu(3\lambda + 2\mu))^{-1}(\lambda + 2\mu)(\lambda^* + 2\mu)\Delta(\bar{\sigma}_{11} + \bar{\sigma}_{22}) + \ldots$$
$$= \mu((-1)^{\alpha+\beta}\partial_{3-\alpha}\partial_{3-\beta}\bar{u}_{3,\alpha}\bar{u}_{3,\beta}) + \ldots,$$

where points denote the different members from (5) if we denote $\bar{\sigma}_{\alpha\beta} = (-1)^{\alpha+\beta}\partial_{3-\alpha}\partial_{3-\beta}\Phi$, from preliminary equation, follows (5) or:

$$\Delta^2\Phi = -0.5E[w,w] + \ldots$$

Now, from St. Venant-Beltrami, compatibility conditions are evident as:

$$\varepsilon_{11,22} - 2\varepsilon_{12,12} + \varepsilon_{22,11} = (2\mu(3\lambda + 2\mu))^{-1}[2(\lambda + \mu)\Delta\bar{\sigma}_{\alpha\alpha} - \lambda\bar{\sigma}_{\alpha\alpha,\alpha\alpha}] - (\mu)^{-1}\bar{\sigma}_{12,12} = 2E^{-1}\Delta^2\Phi,$$

or $\Delta^2\Phi + 0.5E[w,w] \equiv 0$.

The mathematical models considered in [5, ch.I] contain a new quantity, which describes an effect associated with boundary layers. The existence of this member not only explains a set of paradoxes in the two-dimensional elasticity theory (*Babushka, Lukasievicz, Mazia, Saponjan*), but also is very important for the process of generating cracks and holes, for example (see details in [5], ch.1, par. 3.3). Furthermore, let us note that in works [6], equations of the (5)-type are constructed with respect to certain components of stress tensors by differentiation and summation of two differential equations. Also, other equations of the KMR type, which differ from the (5)-type equation, are equivalent to the system, where the order of each equation is not higher than two. For example, in the isotropic case, obviously, for coefficients we have [6]: $c_{\alpha\alpha} = \lambda^* + 2\mu$, $c_{66} = 2\mu$, $c_{12} = \lambda^*$, $c_{\alpha 6} = 0$, $\lambda^* = 2\lambda\mu(\lambda + 2\mu)^{-1}$, λ and μ -are the Lame coefficients. Then, the system (1.7) of [6] is presented in a form:

$$\left(\lambda^{*}+2\mu\right)\partial_{1}\tau+\mu\partial_{2}\omega=\frac{1}{2h}\bar{f}_{1}+\mu(\partial_{1}(\bar{u}_{3,2})-\partial_{2}(\bar{u}_{3,1}\,\bar{u}_{3,2}))-\frac{\lambda}{2h(\lambda+2\mu)}\int_{-h}^{h}\sigma_{33,1}dz\,, \tag{7a}$$

$$-\mu\partial_{1}\omega+\left(\lambda^{*}+2\mu\right)\partial_{2}\tau=\frac{1}{2h}\bar{f}_{2}+\mu(\partial_{2}(\bar{u}_{3,1})-\partial_{1}(\bar{u}_{3,1}\,\bar{u}_{3,2}))-\frac{\lambda}{2h(\lambda+2\mu)}\int_{-h}^{h}\sigma_{33,2}dz, \tag{7b}$$

where the functions: $\tau=\bar{\varepsilon}_{\alpha\alpha}\,,\omega=\bar{u}_{1,2}-\bar{u}_{2,1}$ correspond to the plane expansion and rotation.

For variable thickness of refined theories we have (see details [5], ch.II, point 4):

$$\begin{aligned}
&\frac{1}{h^{3}}D\partial_{\alpha}\left(h^{3}\Delta_{3,\alpha}\left(x_{1},x_{2},\hbar\right)\right)=\sigma_{33}^{+}-\sigma_{33}^{-}-\int_{h_{1}}^{h_{2}}f_{3}dt\\
&+\partial_{\alpha}\left[h\left(\sigma_{\alpha3}^{+}+\sigma_{\alpha3}^{-}\right)-\int_{h_{1}}^{h_{2}}\left(t-\hbar\right)f_{\alpha}dt\right]-\left(h_{2,\alpha}\sigma_{\alpha3}^{+}-h_{1,\alpha}\sigma_{\alpha3}^{-}\right)\\
&+\frac{1}{1-\upsilon}\int_{h_{1}}^{h_{2}}\left(h^{2}-\left(t-\hbar\right)^{2}\right)\Delta f_{3}dt\\
&-\frac{1+\upsilon}{1-\upsilon}\int_{h_{1}}^{h_{2}}\left(h\partial_{\alpha}h+\left(t-\hbar\right)\partial_{\alpha}\hbar\right)f_{3,\alpha}dt\\
&-\frac{h^{2}(2-\upsilon)}{3(1-\upsilon)}\left(1+2\gamma\right)\left(\Delta\sigma_{33}^{+}-\Delta\sigma_{33}^{-}\right)\\
&+\frac{h^{2}}{3}\left[\partial_{\alpha}\left(h+\hbar\right)\Delta\sigma_{\alpha3}^{+}+\partial_{\alpha}\left(h-\hbar\right)\Delta\sigma_{\alpha3}^{-}\right]\\
&+\frac{4h^{2}\partial_{\alpha}h}{3}\Delta\sigma_{\alpha3}\left(x_{1},x_{2},\hbar\right)+R_{1}\left[u_{3}\left(x_{1},x_{2},\hbar\right);\gamma\right]
\end{aligned} \tag{8}$$

where

$$\begin{aligned}
R_{1}\left[u_{3},\gamma\right]=&-\frac{3D}{4h^{3}}\partial_{\alpha}\rho s_{m}\left[\left(h^{2}-\left(t-\hbar\right)^{2}\right)\Delta u_{3,\alpha}\right]\\
&+\rho s_{m}\left[\left(h\partial_{\alpha}h+\left(t-\hbar\right)\partial_{\alpha}\hbar\right)\Delta\sigma_{\alpha3}\right]\\
&-\frac{2-\upsilon}{1-\upsilon}r_{1}\left[\left(t-\hbar\right)\Delta\sigma_{33};\lambda\right].
\end{aligned}$$

Furthermore,

$$\frac{4h}{3}\sigma_{\alpha 3}(x_1,x_2,\hbar)-\frac{2h^3}{3}\Delta\sigma_{\alpha 3}(x_1,x_2,\hbar)$$
$$=-D\Delta u_{3,\alpha}(x_1,x_2,\hbar)+\frac{2}{3}h\left(\overset{+}{\sigma}_{\alpha 3}+\overset{-}{\sigma}_{\alpha 3}\right)$$
$$-\int_{h_1}^{h_2}(t-\hbar)f_\alpha dt-\frac{(1+2\gamma)h^2}{3(1-\upsilon)}\left(\overset{-}{\sigma}_{33,\alpha}-\overset{+}{\sigma}_{33,\alpha}\right)$$
$$+\frac{1+\upsilon}{2(1-\upsilon)}\int_{h_1}^{h_2}\left(h^2-(t-\hbar)^2\right)f_{3,\alpha}dt+R_{1+\alpha}[Q_\alpha;\gamma] \quad (9)$$

(8, 9) expressions without remainder terms R_i present 2D mathematical models of the *Germen-Reissner* type. They give any refined theory choosing an arbitrary parameter γ. For example, if $\gamma=0.1$, we have *Reissner's* type theory; for $\gamma=-0.5$ (7) immediately gives the classical form DE for a deflection equation for variable thickness without any physical or geometrical hypotheses

Using Dirichlet's formula (for repeated integrals, containing v) and the quadrature formula of trapezoid, after some calculations, the last relation will take the form

$$\mu\Delta u_\alpha(x_1,x_2,\hbar)+(\lambda^*+\mu)\partial_\alpha u_{\beta,\beta}(x_1,x_2,\hbar)=f+R_\alpha[u_\alpha]. \quad (10)$$

Here,

$$f_\alpha^h(x_1,x_2)=\frac{1}{2h}\left(\int_{h_1}^{h_1} f_\alpha(x_1,x_2,\hbar)dt-\overset{+}{\sigma}_{\alpha 3}+\overset{-}{\sigma}_{\alpha 3}\right)$$
$$-\frac{\lambda}{2(\lambda+2\mu)}\left(\overset{+}{\sigma}_{33,\alpha}+\overset{-}{\sigma}_{33,\alpha}\right)$$
$$-\mu\left[\Delta\hbar v_\alpha(x_1,x_2,\hbar)+2\partial_\beta\hbar v_{\alpha,\beta}(x_1,x_2,\hbar)\right]-(\lambda^*+\mu)\times$$
$$\left[\partial_{\alpha\beta}\hbar v_\beta(x_1,x_2,\hbar)+\partial_\alpha\hbar v_{\beta,\beta}(x_1,x_2,\hbar)+\partial_\beta\hbar v_{\beta,\alpha}(x_1,x_2,\hbar)\right]$$

$$R_\alpha[u_\alpha(x_1,x_2,\hbar)]=\frac{1}{2h}\left[\mu(\rho_{tr}[(h_1-t)\Delta v_\alpha]+\rho_{tr}[(h_2-t)\Delta v_\alpha])\right]$$
$$+(\lambda^*+\mu)\left[\rho_{tr}[(h_1-t)v_{\beta,\alpha\beta}]+\rho_{tr}[(h_2-t)v_{\beta,\alpha\beta}]\right]-\frac{\lambda}{2h(\lambda+2\mu)}\rho_{tr}[\sigma_{33,\alpha}].$$

We must also remark that v_α unknown functions are defining from preliminary processes, solving the bending problem. The system of DEs (10) without remainder members presents Filon-type equations.

Let us consider *von Kármán-Mindlin-Reissner*-type Refined Theories (RT) in the wide sense when a $\Omega_h = D(x,y)X(h^-,h^+)$ thin-walled, elastic structure is isotropic, the boundary conditions on the $S^\pm$ surfaces have the form: $\sigma_3^\pm = g^\pm$. We remember the explicit form of RT constructions by the generalized Hellinger-Reissner variational principle (see section 2.4 of [5]):

$$\frac{2h^3}{3}\left[\mu\Delta w_\alpha + (\lambda^* + \mu) graddivw_+\right] - \frac{\mu h}{(1+2\gamma)}(w_\alpha + v_{3,\alpha}) =$$

$$(f_\alpha, t) - h(g_\alpha^+ + g_\alpha^-) - \frac{\lambda}{2(\lambda+2\mu)}(\sigma_{33,\alpha}, t)$$

$$\frac{\mu h}{(1+2\gamma)}\left[\Delta v_3 + w_{\alpha,\alpha}\right] = \int_h^{-h} f_3 dt - (g_3^+ - g_3^-).$$

Now, if we will find the σ_3 vector so [5], page 60:

$$\sigma_3 = \frac{(z-h^-)g^+}{2h} + \frac{(h^+ - z)g^-}{2h} + \sum_{s=1}^{\infty} \sigma_3^s(x,y)(P_{s+1}(\frac{z-h^*}{h^+ - h^-}) - P_{s-1}(\frac{z-h^*}{h^+ - h^-})), \tag{11}$$

where $2h = h^+ - h^-, h^* = 0.5(h^+ + h^-), P_s(z)$ are Legendre polynomials. The forms of RT- and *Filon-Kirchhoff* (FK)-type systems of DEs are invariants, and the boundary conditions on the surfaces $S^\pm$ will satisfy exactly for all models. In addition, we remark that:

$$Q_{\alpha 3} = h(g_\alpha^+ + g_\alpha^-) - 2h\sigma_{\alpha 3}^1, M_{\alpha\beta} = \frac{2h^2}{3}\sigma^1{}_{\alpha\beta}, T_{\alpha\beta} = 2h\sigma_{\alpha\beta}^0,$$

$$\int_{-h}^{h} t\sigma_{33} dt = h^2\left(\frac{g_3^+ - g_3^-}{2} - \frac{2}{3}\sigma_{33}^2\right), \tag{12.a}$$

$$\psi_\alpha = \frac{1}{2}\int_{-h}^{h}(h^2 - t^2)\sigma_{\alpha 3} dt = \frac{h^2(1+2\gamma)}{3} Q_{\alpha 3} + r_2\left[t\int_0^t \sigma_{\alpha 3} dt; \gamma\right].$$

$$\int_{-h}^{h} \sigma_{33} dt = h(g_3^+ + g_3^-) - 2h\sigma_{33}^1. \tag{12b}$$

As a reminder, remember $r_2[\,,\,]$ (see (2.16), [5]). It is evident that if we find the solutions of any BVPs for RT and FK, it is possible to define the first and second coefficients of (11). Inversely, if now we use some of the results from [2, 5] when in the systems (6.9) and (6.13) $N = 2$, solutions corresponding with BVPs are given the explicit form for solutions of BVPs for all RTs and FK by (11).

2. To the Approximate Methods of Solving the BVP for Refined Theories

If we consider the BVP: $\Delta u = 0, u\|_{\partial D} = \varphi, D = x^2 + y^2 < R$, we have [7]:

$$u(r,\vartheta) = \frac{1}{2\pi}\int_0^{2\pi} \frac{R^2 - r^2}{R^2 - 2Rr\cos(\vartheta - \lambda) + r^2}\varphi(\lambda)d\lambda .$$

For this simple problem, it evident that the neediness of arithmetic operations for finding an approximate solution (AOS), using any quadrature formula with n points, requests an $O(n^4 \ln n)$ order of Horner units because the calculations of $\cos(\vartheta - \lambda_k)$ by the Fast Fourier Transformation are $O(n \ln n)$ AOS and the number of net points are $O(n^2)$ $-(r_i, \vartheta_j), i = 1,2,...,n; j = 1, n_{1,} n_2,..., n_n$.

This example shows difficulties which arise when we used even the explicit forms of "Potential Theory" for designing and planning.

Here, we must remark, that in the Foreword, when the first edition of this book [1] was published, A. Krilov expressed wishes with respect to practical results. Later, N. Muskhelishvili, A. Gogidze and A. Rukhadze numerically solved some BVPs for Filon-type equations using results from Theories of Analytical Functions, Numerical Integrations and the Method of Least Squares [1, 3]. In [3, pp. 45-48], the Conditional systems of linear algebraic equations as basic system are used. We think that it may be possible to consider the Normal system as well. In this case, the corresponding system is the linear algebraic equation with the square matrix, and it's possible to apply the Fourier Analysis as well.

2.1. Continuous Analogue of the Peaceman-Rechford Alternating Direction Method

A continuous analogue of Peaceman-Rechford's Alternating Direction Method (ADM) [8] and high order accuracy schemes for an approximate solution of von Kármán-Filon-type quasi-linear systems of integro-differential equations were developed.

In the work [9], we prove the theorem associated with the convergence of the iterative process for a strong-elliptic, self-adjoining DE with boundary conditions in $D = [0,1]X[0,1]$:

$$Lu = (k^{ij}\partial_{ij} - q)u(x,y) = -f(x,y),(x,y) \in D,$$
$$u = \varphi(x,y),(x,y) \in \partial D. \tag{1}$$

The continuous analogue of ADM corresponding to (1) defines as the two following serials of expressions:

$$B_1 u^{s+1/2} = B_{12} u^s + f,\ u^{s+1/2}\Big|_{\partial D} = \varphi_1, \tag{2}$$

$$B_2 u^{s+1} = rIu^{s+1/2} + (B_2 - rI)u^s + f, u^{s+1}\Big|_{\partial D} = \varphi_2. \tag{3}$$

Here, r is an iterative parameter, I -unit operator:

$$B_i = \left\|(r+q_i)I - \delta_{\alpha\beta}\partial_{ii} a_i\right\|^n{}_{\alpha,\beta=1},\ B_{12} = B_1 + I, q_1 + q_2 = q, a_i > 0.$$

From this theorem follows that when L is a differential operator of the shell theory of Vekua [2] or Refined Theories of anisotropic elastic plates, in a wide sense, the corresponding results are true. We remark that the problems associated with the existence and solvability for corresponding the BVP of these models and energetic inequalities follow immediately by using corresponding results referenced (e.g., in the monograph) [9].

Now, for same problem, we use the following scheme (when $D = [0,1]X[0,1\}$);

$$(rI - \partial_{11})u^{s+1/2} = (rI + \partial_{22})u^s + f, u^{s+1/2}\Big\|_{\partial D_1} = \varphi_1,$$
$$(rI - \partial_{22}) - \partial_{22})u^{s+1} = (rI + \partial_{11})u^{s+1/2}, u^{s+1}\Big\|_{\partial D_2} = \varphi_2, \partial D_1 \cup \partial D_2 = \partial D. \tag{4}$$

In this case, we have the Lemma (norm by H_{-L} energetically space):

If $\left\|-\partial_{22}u\right\| < \infty, 0 < r < \pi^2$, then for $\forall\varepsilon > 0, \exists S = S(\varepsilon)$ such that when $s > S(\varepsilon), \left\|u - u^s\right\| < \varepsilon$.

Now, for one-dimensional BVPs, if we solve according to the methods of Tikhonov & Samarskii and Volkov or generalized factorization, these processes define the ADM of an arbitrary order exactly with respect to the mesh width. The corresponding AOS for finding the approximate solution has the $O(n^2 \ln n)$ order when $nh = 1$, where h is the mesh width. For the class of 1Dim BVP with a monotone operator, the corresponding results (cited in classical textbooks of *Collatz, Henrici, Keller, Richtmaier. Engel-Miugler & Router, Berezin & Jidkov, Marchuk. Kantorovich & Krilov, Strang & G. Fix, de Boor)* and recent manuals (e.g., of *Quarteroni & Buthcher & Stetter, Bulirsch & Stoer*) may be formulated in the following form: by finite-difference or FEM methods, the approximate solutions converge to the exact solution with no more than a fourth order with respect to the mesh width, AOSs are O(1/ h). Furthermore, the high-order accuracy, three-point schemes were obtained by *Tikhonov & Samarski* and *Volkov*. The constructions of these models contain unstable processes and the orders of AOSs are no less than two because an employment of multipoint formulas of numerical differentiation is necessary for them.

For this case, Stmt.1 is true:

Stmt. 1. *Let us consider the BVP for a linear second order-DE (when the principal part has a selfadjoining form as well or contains small parameters) with Sturm-Liouville boundary conditions. Then, by* $(P),(Q)$ *(see [5]), the expressions constructed are new multi-point stable schemes needing O (1/ h) of the AOS, the convergence of the approximate solution has a* (p-1) *order with respect to h. When p=3, this scheme is identical to classical ones. For 3<p<9, these schemes are different from the Streng & Fix and Mikhlin unstable FEMs.*

2.2. Generalized Factorization Method

According to article [10], the numerical solution of BVPs is below:

$$-(Au+qu)=\frac{d}{dt}(k(t)u'(t))-q(t)\cdot u(t)=f(t), k>0, q\geq 0, 0<t<1, \tag{1}$$

$$u(0)-k_1u'(0)=\alpha,\ u(1)+k_2u'(1)=\beta,\ (k_\alpha\geq 0,) \tag{2}$$

for the method of any order of accuracy; $u(t)$ will be given depending on the order of the smooth of the unknown solution. These numerical schemes were created and published in 1972 [10] and contain a particular case for the corresponding results presented by Marchuk [11, ch.2, point 2.2].

Preliminarily, we shall set the auxiliary formula. They are generalized (P) and (Q) formulae of [5, subsection 13.1]. Thus, we suppose that $u(t)\in C^{p+1}(0,1), p=2s+1$.

(P) formulae have a form (the notation here and below are borrowed from [5, section 13]):

$$u(t_i)=\alpha_i^{p,1}(k)u(t_1)+\beta_i^{p,1}(k)u(t_{i+s})-\sum_{j=2}^{p-1}b_{ij}^{p,1}(k)[Au(t_j)-R_{p-1}(t_i)],$$
$$i=2,3,\ldots,p-1, \tag{3}$$

where

$$\alpha_i^{p,1}(k)=\left(\int_{t_1}^{t_p}k^{-1}(t)dt\right)^{-1}\cdot\int_{t_i}^{t_p}k^{-1}(t)dt,\ \beta_i^{p,1}(k)=1-\alpha_i^{p,1}(k)$$

(Q) formulae are presented as follows:

$$u'(t_i)=\gamma_i^{p,1}(k)[u(t_p)-u(t_1)]+\sum_{j=2}^{p-1}c_{i,j}^{p,1}(k)Au(t_j)+\int_{\tau_1}^{\tau_p}dt\int_{\tau_i}^{t}k(t)Ar_{p-3}(t)dt\,,$$
$$i=1,2,...,p, \tag{4}$$

$$b_{i,j}^{p,1}(\tau)=\frac{1}{\tau_p-\tau_1}\left[(\tau_p-\tau_1)\int_{\tau_1}^{\tau_i}dt\int_{\tau_1}^{t}l_j(t)dt-(\tau_i-\tau_1)\int_{\tau_1}^{\tau_p}dt\int_{\tau_1}^{t}l_j(t)dt\right]$$
$$i,j=2,3,...,p-1,$$
$$c_{i,j}^{p,1}(\tau)=\int_{\tau_1}^{\tau_p}dt\int_{\tau_i}^{t}l_j(t)dt\,,(i=1,2,...,p,\,j=2,3,...,p-1),\ l_j(t)=\prod_{\substack{i=2\\ j\neq i}}^{p-1}\frac{t-t_i}{t_j-t_i}$$

Now, let ω_h designate the net area, as: $\omega_h=\{0=t_1,t_2,...,t_n,t_{n+1}=1;h_i=t_i-t_{i-1}\}$.

As bounding points of the net ω_h, we shall name those t_i knots, for which $i\le s+1$, or $i\ge n-s+1$.

For relation (3) for bounding points it follows that

$$u(t_i)=\alpha_i^{i+s,1}(k)u(0)+\beta_i^{i+s,1}(k)u(t_{i+s})$$
$$-\sum_{j=2}^{2s}b_{i,j}^{i+s,1}(k)k(t_j)Au(t_j)+O(h^{2s+1}),\ i\le s+1; \tag{5}$$

$$u(t_i)=\alpha_i^{n+1,i-s}(k)u(t_{i-s})+\beta_i^{n+1,i-s}(k)u(1)$$
$$-\sum_{j=2}^{2s}b_{i,j}^{n+1,i-s}(k)k(t_j)Au(t_j)+O(h^{2s+1}),\ i\ge n-s+1; \tag{6}$$

The above relations permit to receive expressions of the following form:

$$u(t_i)=\frac{\alpha_i}{1+k_1\gamma_1}[u(0)-k_1u'(0)]+\frac{\beta_i+k_1\gamma_1}{1+k_1\gamma_1}u(t_{i+s})$$
$$-\sum_{j=2}^{2s}\left(b_{i,j}^{i+s,1}(k)-\frac{k_1\alpha_i}{1+k_1\gamma_1}c_{i,j}\right)k(t_j)Au(t_j)+O(h^{2s+1}), \tag{7}$$

$$u(t_i)=\frac{\beta_i}{1+k_2\gamma_{n+1}}\left[u(1)+k_2u'(1)\right]+\frac{\alpha_i+k_2\gamma_{n+1}}{1+k_2\gamma_{n+1}}u(t_{i-s})-$$

$$\sum_{j=n-2s+1}^{n}\left(b_{i,j}^{i+s,1}(k)-\frac{k_2\beta_i\gamma_{n+1}}{1+k_2\gamma_{n+1}}c_{n+1,j}\right)k(t_j)Au(t_j)+O(h^{2s+1}).$$

Here and below, in the coefficients, the top indexes and the dependence of factors on the function $k(t)$ are omitted. In addition, the designation $h=\max_i(t_{i+1}-t_i)$ is entered. A feature of the formulae (7) is that the correct parts contain the same expression, (from conditions (3)), as the data of the initial problem. Obviously, the approach of constructing the formulae of a type (7) allows for the generalization for other conditions.

Let $t_i-t_{i-j}=t_{i+j}-t_i(s+2\le i\le n-s)$. Then the residual member of the formula (3) allows the valuation:

$$\left|\sum_{j=i-s+1}^{i+s-1}b_{i,j}(k)AR_{p-1}(t_j)\right|<c_1M_{p+1}h^{p+1},\ M_{p+1}=\max_{(0,1)}\left|u^{(p+1)}(t)\right|.$$

For interior knots $t_i\in\omega_h$ from expression (3) there follows:

$$u(t_i)=\alpha_iu(t_{i-s})+\beta_iu(t_{i+s})-\sum_{j=i-s+1}^{i+s-1}b_{i,j}(k)Au)+O(h^{2s+2}) \tag{8}$$

If now in the formulae (8) we replace the expression Au by $qu+f$ and then omit the remainder term, we shall obtain algebraic system of linear equations, the solution of which shall designate through u_i, $(i=2,3,...,n)$.

The matrix appropriate to this system is a multi-diagonal matrix depending on s. For the solution of such systems, the classical factorization method is easily applied.

A system of equations concerning the values u_i, received from (7), for convenience we shall rewrite as:

$$u_i=\frac{\beta_i+k_1\gamma_1}{1+k_1\gamma_1}u_{i+s}+\sum_{j=2}^{2s}d_{ij}u_j+F_i, i=2,3,...,s+1, \tag{9a}$$

$$u_i=\alpha_iu_{i-s}+\beta_iu_{i+s}+\sum_{j=i-s+1}^{i+s-1}d_{ij}u_j+F_j, i=s+2,...,n-s \tag{9b}$$

$$u_i=\frac{\alpha_i+k_2\gamma_{n+1}}{1+k_2\gamma_{n+1}}u_{i-s}+\sum_{j=n-2s+1}^{n}d_{ij}u_j+F_i,\ i=n-s+1,...,n \tag{9c}$$

where, for example,

$$F_i = \frac{\alpha_i}{1 + k_1\gamma_{1,i}}\alpha + \sum_{j=2}^{2s} d_{ij} f(t_j), i \leq s+1.$$

The first s of the formulae give the following recurrence expression:

$$u_i = A_i u_{i+s} + \sum_{\substack{j=i+1 \\ j \neq i+s}}^{2s} A_{ij} u_j + B_i \,, i = 2,3,\ldots,s+1, \tag{10}$$

where

$$A_{ij} = \frac{e_{ij}}{1 - e_{ij}}, \; j = i+1,\ldots,2s, \; j \neq i+s,$$

$$A_i = A_{i,i+s} = \frac{\beta_i + k_1\gamma_1}{(1 - e_{ii})(1 + k_1\gamma_1)} \tag{11}$$

$$e_{ii} = d_{ij} + \sum_{k=2}^{i-1} d_{ik} \sum_{l=k}^{i-1} A_{lj} \prod_{m=k}^{l-1} A_{m,m+1}, \; \prod_{m=k}^{l-1} \cdot = 1, \; k > l-1,$$

$$B_i = \frac{F_i + \sum_{k=2}^{i-1} d_{ik} \sum_{l=k}^{i-1} B_l \prod_{m=k}^{l-1} A_{m,m+1}}{1 - e_{ii}}, \; i = 2,3,\ldots,s+1.$$

Let i be the number of any internal point of the net area ω_h. Then, from expressions (9-11) follows:

$$u_i = A_i u_{i+s} + \sum_{j=i+1}^{i+s-1} A_{ij} u_j + B_j \,, \; i = s+2,\ldots,n-s, \tag{12}$$

where

$$A_{ij} = \frac{e_{ij}}{1 - e_{ij}}, \; j = i+1,\ldots,2s, \; j \neq i+s,$$

$$A_i = A_{i,i+s} = \frac{\beta_i}{(1 - e_{ii})}$$

$$e_{ii} = d_{ij} + \sum_{k=i-s+1}^{i-1} d_{ik} \sum_{l=k}^{i-1} A_{lj} \prod_{m=k}^{l-1} A_{m,m+1} + d_i A_{i-s,j}, \tag{13}$$

$$B_i = \frac{F_i + \sum_{k=i-s+1}^{i-1} d_{ik} \sum_{l=k}^{i-1} B_l \prod_{m=k}^{l-1} A_{m,m+1} + d_i B_{i-s}}{1-e_{ii}}, \; i = 2,3,....,s+1.$$

The values $u_i, i = n-s+1,...,n$ satisfy the following equalities:

$$u_i = \sum_{j=i+1}^{i+s-1} A_{ij} u_j + B_i \,, i = n-s+1,...,n-1, \tag{14}$$

where

$$A_{ij} = \frac{e_{ij}}{1-e_{ij}}, \; A_i = A_{i,i+s} = \frac{\beta_i}{(1-e_{ii})} \tag{15}$$

$$e_{ii} = d_{ij} + \sum_{k=i-s+1}^{i-1} d_{ik} \sum_{l=k}^{i-1} A_{lj} \prod_{m=k}^{l-1} A_{m,m+1} + \frac{\alpha_i + k_2\gamma_{n+1}}{1+k_2\gamma_{n+1}} A_{i-s,j},$$

$$B_i = \frac{F_i + \sum_{k=i-s+1}^{i-1} d_{ik} \sum_{l=k}^{i-1} B_l \prod_{m=k}^{l-1} A_{m,m+1} + \frac{\alpha_2 + k_2\gamma_{n+1}}{1+k_2\gamma_{n+1}} A_{i-s,j}}{1-e_{ii}}.$$

At last, the value u_n defines explicitly:

$$u_n = B_n, \tag{16}$$

$$B_n = \frac{F_n + \frac{\alpha_n + k_2\gamma_{n+1}}{1+k_2\gamma_{n+1}} B_{n-s} + \sum_{k=n-s+1}^{n-1} d_{nk} \sum_{l=k}^{n-1} B_l \prod_{m=k}^{l-1} A_{m,m+1}}{1-e_{ii}},$$

$$e_{nn} = d_{nn} + \sum_{k=n-s+1}^{n-1} d_{nk} \sum_{l=k}^{n-1} A_{\ln} \prod_{m=k}^{l-1} A_{m,m+1} + \frac{\alpha_n + k_2\gamma_{n+1}}{1+k_2\gamma_{n+1}} A_{n-s,n} \tag{17}$$

Let α_i and β_i satisfy the following bilateral inequalities:

$$\frac{1}{s} < \beta_i, \; \alpha_{n+1-i} < \frac{1}{2}, i = 2,3,...,s+1, \; 1 - c_3 h^2 < \alpha_i \beta_i^{-1} < 1 + c_4 h^2,$$
$$c_3, c_4 > 0, \; i = s+2,...,n-s,$$

where c_3, c_4 are constants. Obviously, it is possible by the appropriate choice h (see expressions for α and β from the formula (3)).

Then from (11), (13) and (15) follows:

$$A_{ij} < (1 - c_5 h)\max_{i \le s+1}\{A_i, A_{s+2}\}, B_i < c_6\alpha + c_7\beta + c_8 \max_j |f(t_j)|, \tag{18}$$

where non-negative constants c_5, c_6, c_7, c_8 do not depend on h.

The conditions (18) are the definition of stability of the computing process by formulae (11), (13) and (15) concerning the initial data and correct part accordingly.

The stability of process (10), (12) and (14), for calculating u_i values is also obvious, as the operator appropriate for these expressions is an compression operator. From the above stated formulae it follows that the method of generalized factorization is optimum, as the number of arithmetic operations necessary for calculating the approximate solution u_i is directly proportional to the number of points of the net areas ω_h where

$$\sigma'_{(k-1)z+i} = \frac{2}{k_1 + k_2 + k_1 k_2}\left\{ k_1\left[\frac{1}{2}(2 + k_2)A_k + \sum_{i=2}^{k-1}(2k + ik_2)A_{2k-i} + (k + k_2)A_{2k-1}\right]\right.$$
$$\left. - k_2\left[\frac{1}{2}k(2 + k_1)A_k + \sum_{i=2}^{k-1}(2k + ik_1)A_i + (k + k_1)A_1\right]\right\} - k\sum_{j=2}^{2z} c_{i,j} y''_{(k-1)z+j} + O(h^{p-1}).$$

The construction of the one-parameter schemes will be completed, if we attach two Cauchy (initial) problems to the expressions of types (6) and (7):

$$u_1'(x) = f(x, \lambda(x), u_1(x)),\ l_1 \le x \le 1,$$
$$u_1(l_1) = \gamma,\ l_1 = x_{(k-1)z+1},$$
$$u_1'(x) = f(x, \mu(x), u_1(x)),\ l_2 \ge x \ge 0,$$
$$u_1(l_2) = \delta,\ l_2 = x_{(k+1)z+1}.$$

Now, we return to study the problem (1)-(2) and introduce the following values:

$$\omega_1 = \frac{1}{8} + \frac{1}{4(k_1 + k_2 + k_1 k_2)}\left(4 + k_1 + k_2 + \frac{(k_2 - k_1)^2}{k_1 + k_2 + k_1 k_2}\right),$$
$$\omega_2 = \frac{1}{2(k_1 + k_2 + k_1 k_2)}(k_1 k_2 + 2\max\{k_1, k_2\}), \tag{19}$$
$$\omega_2' = \frac{1}{2} - \frac{k_1 k_2}{4(k_1 + k_2 + k_1 k_2)},\quad \omega = \max\{\omega_2, \omega_2'\}.$$

The following theorem is true:

Theorem 1. *Let the function* $f(x,u(x),u'(x))$ *be continuous with respect to* x, *satisfying a Lipschitz's condition relative to* u *and* u' *with constant* L *and* L' *respectively; in addition, let one of two conditions be executed:*

$$\omega(L+L')<1,\omega_1 L+\omega_2 L'<1. \tag{20}$$

At these restrictions, the initial problem has a unique solution which can be constructed by an iterative method.

The proof of this theorem coincides with the scheme of the proofs of theorems 13.2 and 13.3 [5].

Now, in the formulae of types (6) and (7), we omit the remainder terms. We get the expressions for constructing the initial table. The Cauchy problem we shall replace by the multistage methods. We shall name the resulting system as the difference scheme.

The following theorem is true:

Theorem 2. *For the problem (1)-(2) let one of conditions (9) be true. Then:*

1) *the difference scheme has a unique solution and the iteration method converges;*
2) *as in the case of the uniform grid* $(p=3,5,7)$, *as in the case of Gaussian grid* $(p>3)$, *convergence of the algebraic analogue's solution to the solution of a problem (1)-(2) and its derivative has a* $(p-1)$*-degree respect* h.

The proof of this theorem is similar to the proof of the theorem 13.2 [5].

The following theorem is true:

Theorem 3. *The number of arithmetic operations which is necessary for the calculation of an approximate solution* $\overline{u}(x)$ *and its derivative* $\overline{u'}(x)$ *has the order* $k\cdot\ln k$.

The proof of this theorem is based on the specific character of sums σ_{tz+1}. If we calculate σ_{kz+1}, then σ_{tz+1} $\forall t\neq k$ will be calculated, as it is contained in σ_{kz+1} as subsums.

The practical convenience of a generalized factorization method constructing an algebraic analogue is associated with the following: the difference from the other high-accuracy methods of Tichonov, Samarski and Volkov is not necessary to be present to make up the table of multiple integrals or derivatives from initial data.

References

[1] N. Muskhelisvili. *Some Basic Problems of Mathematical Theory of Elasticity*, M.: Nauka, 1966.

[2] I. Vekua. *Shell Theory: General Methods of Construction*, Pitman Advance Publ. Prog., Berlin/London/Montreal, 1985.

[3] A. Gorgidze. *Some Generalizations of Bending and Twisting Problems Compound Beams*, Tbilisi: Metsniereba, 2001. At consideration of schemes having arbitrary order of accuracy we will study also the processes connected with their numerical realizations.

[4] P. Ciarlet *Mathematical Elasticity: II, Theory of Plates.* Elsevier: Amsterdam /London/ New-York, 1997.

[5] T. Vashakmadze. *The Theory of Anisotropic Elastic Plates,* Kluwer Acad. Publ & Springer. Dortrecht /Boston/ London, 2010.

[6] T. Vashakmadze. On the Basic Systems of Equations of Continuum Mechanics and Some Mathematical Problems for Anosotropisc Thin-walled Structures, IUTAM Symposium on Relations of Shell, Plate, Beam and 3D Model, dedicated to the Centerary of Ilia Vekua's Birth (Edited by G. Jaiani, P. Podio-Guidugli), *Springer Science Business Media B*. V.9, 2008, 207-217.

[7] L. Kantorovich, V. Krilov. *Approximate Methods of High Analysis*, M. Fizmatgiz, 1962.

[8] D.W. Peaceman, H. H. Rachford, 1955, The numerical solution Parabolic and elliptic DEs, *J. Soc. Ind. Appl. Math*, 3, 1955, 28-41.

[9] N. Morozov. *Selected two-dimensional problems of the theory of elasticity*, St. Peterburg University Press, 1978.

[10] T. Vashakmadze. Methods of generalized factorization and its applications to numerical solution some probl. mathem. physics. *Proceed. Symp. Continuum Mechanics and Related Problems of Analysis*. Tbilisi:Mecniereba, v.I, 36-45

[11] G. Marchuk. *Methods of Numerical Mathematics.* M.: Nauka, 1989.

In: Several Problems of Applied Mathematics and Mechanics ISBN: 978-1-62081-603-5
Editors: Ivane Gorgidze and Tamar Lominadze

Chapter 2

COGNITIVE APPROACH TO THE RESEARCH AND MODELING OF THE GEORGIAN FUEL AND ENERGY COMPLEX

Iuza Sh. Verulava, Archil I. Prangishvili, Zurab A. Gasitashvili[1] and Irakli D. Rodonaia
Georgian Technical University, Tbilisi, Georgia

Abstract

Important problem areas of fuel and energy complex based on the cognitive approach have been determined in the paper. The system analysis, taking into consideration strong and weak properties, risks and chances, environmental, political, legal, social, economic, psychological and other factors, has been conducted. The methodology of constructing cognitive models is proposed. As a result of modeling, the possible scenarios of development of the energy economy have been investigated.

Model Representation of the Fuel and Energy Complex

There are various ways of studying and interpreting realistic situation in the Georgian energy economy. For example, the development of the energy economy may be considered in the aspects of political decisions, macroeconomic rules and trends, or technological issues. At the same time psychological factors such as private interests, workers' motivation, etc. become very important.

Decision making in management of the energy economy is difficult due to the following factors:

[1] E-mail address: zur_gas@gtu.ge

- Processes being developed inside the energy economy complex are closely interrelated, and therefore, it is impossible to choose a single aspect and study it in detail; all processes inside the system must be studied in complex
- Lack of quantitative information about process dynamics, which forces the qualitative data to describe these processes
- Time-varying (transient) nature of processes, at the same time the nature of change of some process characteristics are often unknown, which makes it difficult to build their quantitative models

The above-mentioned peculiarities are reason to refer to such systems as **weakly structured** ones. Study of these systems by traditional econometric methods is often impossible. The alternative is to apply cognitive structuring of processes and generating of corresponding decisions.

The cognitive analysis implies the cause-and-effect structuring of process information according to the following scheme [2]:

1. Any event in the system is caused by the movement of material (resources, money, etc.) and non-material (information) flows. Any movement of a flow can be described by a chain of cause-and-effect relations, which represents an analyst's knowledge about processes in the system; the revealing of such flows is the first step in the cognitive analysis of the system under study.
2. Each flow from the selected ones is described by the relevant factors; their integration represent the factor set, in terms of which the process being developed in the system are described.
3. Interaction of factors is considered in terms of cause-and-effect chains describing the movement of each flow.

The strength of the reciprocal influence of factors is described by such linguistic variables as "significant," "normal," "weak," etc.

The reciprocal influence of factors is represented by a so-called cognitive map, which is, in fact, weighted organizational graph of a system being investigated.

When building a cognitive map the following methods may be applied:

1. The analyst builds the cognitive map based only on his knowledge, without using experts' knowledge and other information; the efficiency of this method depends on the qualification of the analyst and his ability to determine the interrelation between basic system factors. It seems to be the fastest way of building cognitive maps, which at the same time provides confidence.
2. The analyst builds the cognitive maps based on expert documentation. The advantage of this approach is its flexibility, but, on the other hand, it is rather labor-intensive and requires co-ordination of the experts' knowledge.
3. The analyst builds the cognitive map with the help of a group of experts; the advantage of this method is the aggregation of individual knowledge.
4. The analyst builds the cognitive map based on open polls; its advantage is an active dialog with sources of information.

Graphic models of weakly structured systems are built with two purposes:

- More detailed structuring of the system under study
- Using structuring to solve the initial problems within the system under study.

We call factors that represent the essence of the system behavior as objective factors. The control factors (model's "input" factors) are those factors that the analyst can use to control the system processes by applying certain influence signals.

Model's external factors. They are the model graph's hanging nodes. They correspond to those environmental factors whose dynamics do not depend on the state of the weakly structured system. Hence, the external factors are not devoted to controlling the system but reflecting the influence of the environment on the system being modeled.

The problem of the controlling model's dynamic consists of the selection of a set of such control impacts that compensates for the undesirable results of initial conditions and provide the necessary dynamics of change of the objective factors. When modeling the weakly structured system, it is necessary to pass from the graphic model to the time equations when in these equations is represented the graph's structural properties. The computer realization of the model of such systems implies the initial description in the discrete time intervals.

The initial state of a model is a unity of changes of directions and speeds of all factors in an initial moment of time.

Objectives of control (objective vector) is a vector of objective factors, given values of changes of all their coordinates.

Self-development of model's processes. Before solving the problem of control, which provides the desired development of processes in the system, it is necessary to analyze the free development of processes (without control impacts) from a given initial state. The structure of the model and initial conditions may happen to be such that the free self-development of the system will lead to the desired objective state, i.e., the tendency of changes of objective factors will approach the desired change tendency. If objective factors move away from the desired tendency, then the correction of objective factors by yielding the correspondent control impact is required.

Principles of Determination of Cognitive Modeling Factors and Interrelations Between Them

The energy economy as a production system operates in a certain environment and interrelates with its various subjects, such as the government, regions, international partners, state and commercial organizations, etc. The environment changes its state and conditions, which creates additional difficulties and risks to the energy economy. For purposeful development of the energy economy it is necessary to develop new tools to represent such difficulties and risks. When studying the problem situation, the great importance also has a human element (in this case the management of the production system), that is, the internal understanding of current situation in the system (the picture "Model of the World"). The "Model of the World" contains a set of grounds, perception of special valuable and practical principles, which the subject uses in his managerial activity of generating impacts for development of situations.

When constructing the “Model of the World,” not only the “picture of the present” is created, but the future of the development of the energy economy is considered as well. Hence, the “preventive” research may be used for creation of the space of possibilities, which we can use in one of the methods of perception - modeling.

The cognitive modeling, which is the continuation of representing processes, allows us to explore the space of opportunities of the energy economy development. Using the cognitive modeling the “Compact Model of the World” can be constructed. Here non-significant factors are isolated and are excluded from consideration.

When determining the basic factors the main criteria are:

- Mechanisms that support the purposeful development of the energy economy in the conditions of interactions with the environment
- Interests and expectations of participants of the aroused problem situation

In the cognitive modeling each factor is determined by its influence on other factors. This type of interaction allows the interests and expectations of participants to be taken into account.

The interaction between factors can be positive, when growth (reduction) of a factor causes growth (reduction) of other related factors, or negative, when growth (reduction) of a factor causes reduction (growth) of other related factors.

As a result the development of the energy economy is represented by interactions of basic factors of the so called cognitive map, whose nodes are unambiguously associated with factors and arcs represent the connections between factors.

Development of Model of Interaction Between Energy Sector with Other State Subjects

The development of this model begins with the building of the cognitive map that represents the real situation. Various factors of development of the energy economy can be considered: for example, political decisions or macro economic rules and trends, or social conditions in the society. On the basis of the built cognitive map the self-development of the system is being modeled. The goal is to reveal the positive tendency in the development of the energy economy. With the aid of cognitive modeling it is possible to fix the evolution of the situation consisting of two components:

- Self-development of the situation, that is, the free development
- Control of development of the situation determined by the selected control objective

Factors of the Cognitive Factors That Characterize the Situation "Today"

In this situation the following problems can be selected:

- In the control of the fuel and energy complex the strategic interests of the country are not determined
- Extremely hard situation in the fuel and energy complex from the point of security
- Financial state of the fuel and energy complex is anxious

All above-mentioned aspects create many unsolved problems for the fuel and energy complex. These problems are very critical for the country because they generate social tension in the society and reduction of living standards of population.

Internal Factors of the Energy Economy

Financial and production activity of the fuel and energy complex is based on the following: the fuel and energy complex at the expense of rather modest turnover means received energy. Then the complex makes the realization of energy to customers. They are generation objects, energy transmission, distributive companies, wholesale market, and dispatching services. Each of them represents an independent juridical person. For their normal functioning and reproduction the full financial flows are necessary. The analysis shows that many factors influence the process: commercial losses, technical losses (above the standard), technical failures, lack of turnover means, increased expenses, low collection of payments for energy consumed, etc.

External Factors of the Energy Economy.

The successful functioning of the fuel and energy complex is impossible without the impacts from the outer world (social, economic, political, etc). On cognitive maps these impacts (direct or indirect) can be expressed by means of a "boundary layer" and "outer field."

In the completed model the external factors, which represent other subjects of the country and governmental influence on the fuel and energy complex, are very important not only during the analysis of the current situation but for the prediction of the behavior of the system. These factors are the social situation in the country, economic situation, tax situation, political situation, etc.

Cognitive Map of a "Boundary Layer"

Determination of Degree of Reciprocal Influence of Factors Characterizing the Situation "Today"

The unity of an internal field, a boundary layer and an external field creates the complete field of factors of the fuel and energy complex. In an interdependency matrix immediate relations between factors are presented. The columns and rows of the matrix correspond to the factors of the cognitive map, and numerical values at the intersection of a*i*-th row and a *j*-th column show those weights and direction by which the *i*-th row influences on the factor of the *j*-th column. To describe the reciprocal influence of factors the set of linguistic variables, such as "strong," "moderate," "weak," etc., is used. The numerical values of linguistic variables are reflected on the closed interval [0, 1] as follows: "very weak" – 0.1, "moderate" – 0.3, "significant"- 0.5, "strong"- 0.7, and "very strong"- 1.0. The direction of influence is determined by its sign: a positive one, when growth (reduction) of a factor will also cause growth (reduction) of another related factor, and a negative sign, when growth (reduction) of a factor will cause reduction (growth) of another factor. For example, the moderate (-0.3) reduction of the second factor, "tax benefits of the fuel and energy complex," causes the growth of first factor, "tax values." That is, the greater the tax benefits, the fewer the taxes, the greater the "profit," the greater the "turnover means." This relation is reflected by the values of coefficients 0.2 and so on (filling matrices is done by experts).

Modeling of Self-development Scenario in the Energy Economy (Scenario "Today")

This scenario is based on observation of changes in the system (in fact, it is the task of exploitation). It is possible to obtain a situation that satisfies the requirements of development of the fuel and energy complex. Then, the conclusion is the intervention in the process of operating the system. These results of modeling, in fact, answer the question, "What can happen in the future, if 'today's' tendency will hold out?" In this case the following must be entered in the model:

- interdependency matrix
- initial values of factors

The initial tendencies represent the current situation in the fuel and energy complex. They are given in the form of linguistic variables.

The concrete values of linguistic variables are given according to the poll of experts in 10-point scale and 3 – positional tendency scale: growing "+1," unchanged "0" and decreasing "-1."

The First Scenario –"Today"

The starting data of the model (values of the factors) are selected on the basis of the current crisis in the energy. The model corresponds to the self-development scenario.

The results of modeling are given below. This situation is represented by the first scenario. The factors to be observed or objective functions and their predictions are given on the graph. The yellow colored rectangles show the initial tendencies; the red colored ones show objective tendencies that correspond to the desired development of the situation, and blue colored ones, the predicted results tendencies.

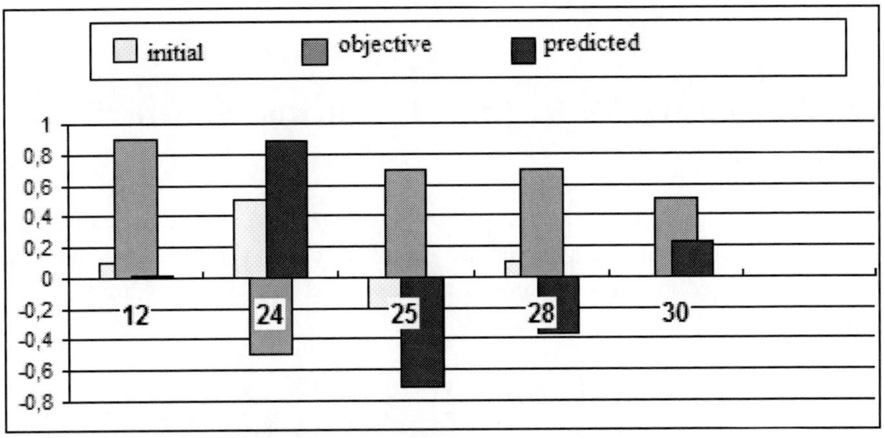

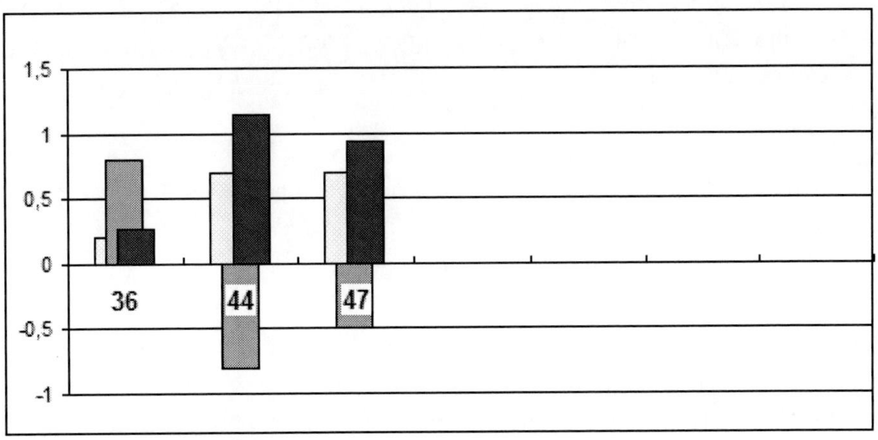

12. Profit from sales does not change
24. Shadow turnover significantly increased
25. Efficiency of the fuel and energy complex significantly decreased
28. Ecological security moderately decreased
30. Rehabilitation slightly increased
44. Corruption strongly increased
47. Social tension strongly increased

Graph of the results.

As the results show, in the condition of self-development the situation was not improved, but, on the contrary, became significantly worse and became the source of new problems. In particular, the 12-th factor is the collection of payment for the consumed energy, and it decreased compared with the "today" collection(it is contrary to desired dynamics of changes); the 24-th factor, the shadow turnover, is increased; the 25-th factor, the efficiency, is decreased; the 28-th factor, ecological security, is decreased; the speed of change of the 30-th factor is insufficient; and the 44-th factor, the level of corruption, is increased.

Conclusions. All the above makes it necessary to intervene radically in the operation of the fuel and energy complex, to change significantly the strategy and tactics of its development. That implies the use of the second modeling option (modeling of controlling development of the fuel and energy complex)

Modeling Using Control (On the Basis of the Starting Scenario "Today Situation")

The controlling factors from the list of existing ones were selected. Predictions for not only variants were obtained but also for the set of accepted variants, each determined by concrete values of controlling parameters (these are selected according to experts' knowledge).

The factors selected as controlling ones:

- 2 - Tax benefits in the fuel and energy complex
- 6 - Turnover means of the fuel and energy complex
- 8 -Commercial losses
- 12 - Profit from realization
- 17 -Tariff on electric power
- 30 - Ecological security in the fuel and energy complex
- 33 -State order
- 34 -Level of participation in transnational unions
- 37 -Demand on electric power
- 43 -State grants for the fuel and energy complex
- 49 -Export of peak
- 50 - Level of legislative base in the fuel and energy complex

- The factors 2, 6, 8, 12, 17, 33, 34, 43, 50 provide the governmental control.
- The 30-th controlling factor provides actions according to the Kioto protocol.
- Factors 39 and 49 represent actions reflecting participation of the Georgian fuel and energy complex in transnational unions.

The starting data of the second scenario correspond to the "today" situation. The model corresponds to the control scenario (parameters to regulate crisis situation are determined).

The results of modeling (the second scenario) are shown below:

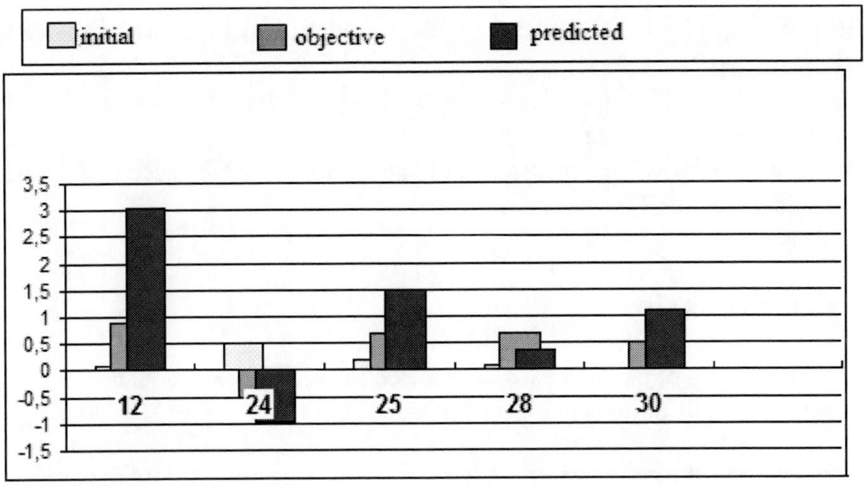

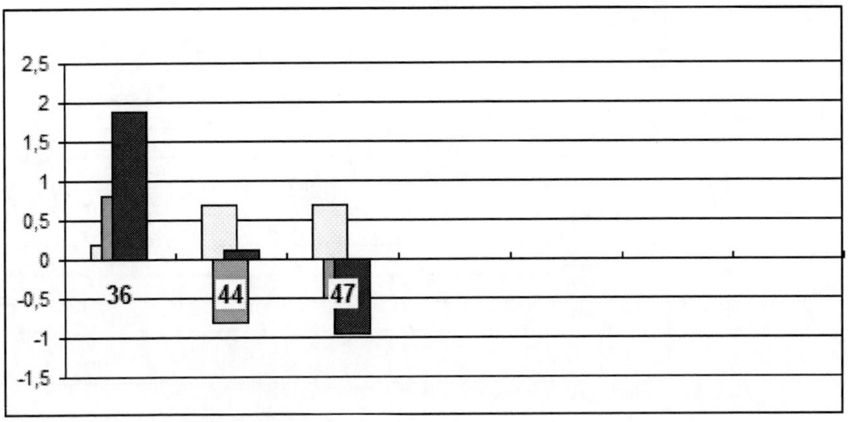

12. Profit from sales strongly increased
24. Shadow turnover strongly increased
25. Efficiency of the fuel and energy complex strongly increased
28. Ecological security moderately decreased
30. Rehabilitation strongly increased
44. Corruption strongly increased
47. Social tension strongly increased

Conclusions. The second scenario shows that Georgia has to participate more intensively in various transnational unions. It shows also that the governmental intervention must be strengthened (of course, if it is compatible with the market economy principles and if there is analogue in the world practice). The membership of Georgia in the international energy system guarantees that the intensification of the governmental control is not contrary to the market economy principles. Besides, the intensification of the governmental control is based on the famous method of market economy organization - the Keyenes theory [3], which has proved its efficiency in many developed countries.

Modeling of the Self-development in the Good Starting Conditions (Scenario "Tomorrow")

For this scenario modeling of self-development of the fuel and energy complex was conducted. The relevant values of factors and tendencies of stable successful countries were selected as initial factors.

The Third Scenario

The starting values of parameters were selected in accordance with their desirability. The graph of results of modeling is given below:

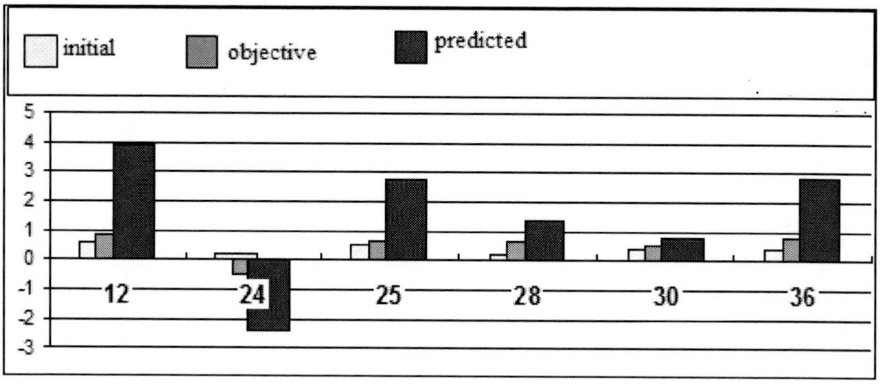

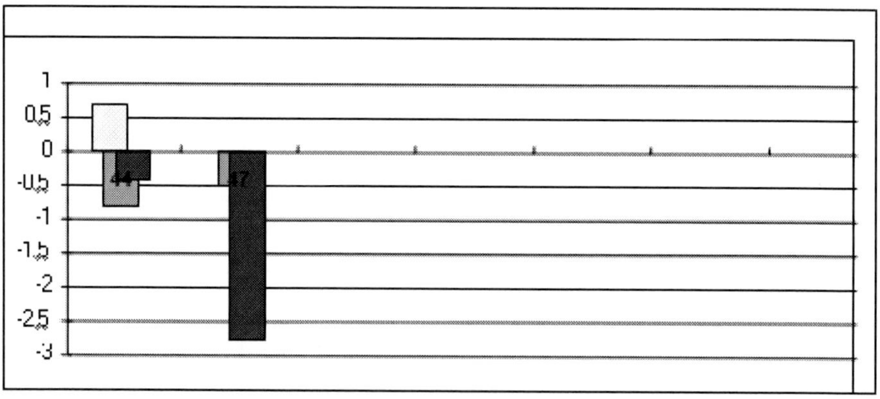

12. Profit from sales strongly increased
24. Shadow turnover strongly decreased
25. Efficiency of the fuel and energy complex strongly increased
28. Ecological security moderately decreased
30. Rehabilitation strongly increased
44. Corruption moderately decreased
47. Social tension strongly decreased

As one can see, the results are very good, and their economic action principles are based on monetary development theory.

References

[1] Gasitashvili Z., Verulava I., Verulava D., Prangishvili A.,Cognitive approach to the research and modeling of the Georgian fuel and energy complex. *Electronic journal*: http://gesj.internet-academy.org.ge, №1(8), 2006.

[2] Alekseev N.A., *Spontaneous phenomena in nature: manifestation, effectiveness of protection*, Moscow., 1998.

[3] Prangishvili I.V., Ostrovsky U.I., Monetarism and Keynesian economics in modern Russia's environment, *Proceedings of Control Problems*, Institute of Russian Academy of Science, Moscow,2002.

[4] *Computer systems of modeling of cognitive cards: approaches and methods*, Kulinich A.A., Control Science #3, 2010.

In: Several Problems of Applied Mathematics and Mechanics ISBN: 978-1-62081-603-5
Editors: Ivane Gorgidze and Tamar Lominadze

Chapter 3

SPINORS' METHODS IN ROBOTS SPATIAL MOVEMENT TERMINAL CONTROL

A. Prangishvili and A. Milnikov
Georgian Technical University, Tbilisi, Georgia

Abstract

In the article, using spinor representation of orthogonal transformations, the expressions between second order complex unitary transformations matrixes and real orthogonal matrixes of spatial rotations in three dimensional Euclidean space L^3 are received, which allows easily calculating of corresponding Euler's angles. The obtained results have enabled reducing the actually three-dimensional problem of spatial motion control to the one-dimensional problem; control kinematical functions of Euler's angles and control spinor matrix of rotation were constructed, by means of which the control process of spatial rotations is completely determined.

Keywords: control spinor matrix, hermitian functionals; spatial rotations; terminal control

1. Problem Formulation

Methods of representation of three-dimensional rotations used in the solving of various engineering problems are usually confined to the description of individual concrete rotations centered at the origin (zero center). Among these methods is, in particular, the well known method of orthogonal real matrices, whose elements are functions of Euler angles [1, 2]. At the same time it should be said that the problem of describing so-called generalized rotations [3] evokes a much greater interest both from the theoretical standpoint and from the standpoint of applications (in the first place we mean an application in robotics and, in particular, in the planning of trajectories in the case of obstacles). Under generalized rotations we mean the set of all possible rotations with both zero and nonzero centers which transform the initial three-dimensional point to the finite one. The basic problem arising in this context can be formulated as follows: given two three-dimensional points $x(x^1, x^2, x^3)$ and

$y(y^1, y^2, y^3)$, it is required to define the set of all possible transformations and centers of rotations that bring about the transformation of the point x to the point y. It is obvious that this problem can be easily extended to the case where instead of two points we consider two finite sets of points, $\{x_i\ (x_i^1, x_i^2, x_i^3\}$ and $\{y_i\ (y_i^1, y_i^2, y_i^3\}$ i=1,2,...m, which corresponds to the case of rotations of a solid.

2. Problem Solution

2.1. Spinorial Representation of Spatial Rotations

Let L^3 be a linear Euclidean space with orthonormalized basis vectors e_1, e_2, e_3. To each vector $x = x^1 e_1 + x^2 e_2 + x^3 e_3$ of the space L^3 we assign a traceless Hermitian matrix

$$X = \begin{vmatrix} x^3 & x^1 - ix^2 \\ x^1 + ix^2 & -x^3 \end{vmatrix}, \tag{1}$$

whose elements are the so-called spinor components of the vector x. When we pass from the usual Euclidean components of the vector x to the spinor ones, we thereby identify the vector x with Hermitian functionals on the two-dimensional linear space C^2 over the field of complex numbers C [4]. Denote by $L(C^2)$ the set of all Hermitian functionals on C^2 and consider the following decomposition

$$X = x^1\sigma_1 + x^2\sigma_2 + x^3\sigma_3, \tag{2}$$

where

$$\sigma_1 = \begin{vmatrix} 0 & 1 \\ 1 & 0 \end{vmatrix}, \sigma_2 = \begin{vmatrix} 0 & -i \\ i & 0 \end{vmatrix}, \sigma_3 = \begin{vmatrix} 1 & 0 \\ 0 & -1 \end{vmatrix}$$

are Pauli matrices [4].

From decomposition (2) it follows that the set $L(C^2)$ is a linear three-dimensional space over the field of real numbers and, thus, it can be identified with L^3. Note that to each basis vector of the two-dimensional space C^2 we can assign the basis elements $\sigma_1, \sigma_2, \sigma_3$ of the space $L(C^2)$ (and also the orthonormalized basis vector e_1, e_2, e_3 due to the identification of L^3 and $L(C^2)$): each of the matrices σ_i is represented as some linear combination of tensor products of basis vectors of the space C^2 [4]. The foregoing reasoning implies that for any matrix $C \in C^2$, which is a matrix of transformation between two basis vectors of the space C^2, there also exists a transformation matrix of the corresponding orthonormalized basis vectors in the space L^3.

Proposition. The matrix of transformation of the basis elements in C^2 is unitary.

Proof. If on the space C^2 we consider Hermitian functionals of the form

$$X = \begin{vmatrix} x^0 + x^3 & x^1 - ix^2 \\ x^1 + ix^2 & x^o - x^3 \end{vmatrix},$$

then they will correspond to the four-dimensional vectors of a pseudo-Euclidean space with signature (1,3) and with basis vectors

$$\sigma_0 = \begin{vmatrix} 1 & 0 \\ 0 & 1 \end{vmatrix}; \sigma_1 = \begin{vmatrix} 0 & 1 \\ 1 & 0 \end{vmatrix}, \sigma_2 = \begin{vmatrix} 0 & -i \\ i & 0 \end{vmatrix}, \sigma_3 = \begin{vmatrix} 1 & 0 \\ 0 & -1 \end{vmatrix}. \quad (3)$$

Now, transformations of the basis vectors of the two-dimensional space C^2 lead to transformations of the basis vectors (3), while the transformation matrices remain the same as in the case of functionals of form (1). The orthogonal complement $\perp\sigma_0$ of the first basis vector σ_0 is an anti-Euclidean space (because of the pseudo-Euclidean property of the space defined by vectors (3)) and, after changing the signs of the scalar products, a three-dimensional Euclidean space that coincides with $L(C^2)$. The narrowing of the action of matrices of basis vector transformation in C^2 to the subspace $\perp\sigma_0$ means that these matrices satisfy the condition $\overline{C}^T \sigma_0 C = \sigma_0$, i.e. $C^{-1} = \overline{C}^T$, Q.E.D.

The problem posed in Subsection 1 can be now reformulated in terms of the spinor space C^2: Given two traceless matrices of Hermitian functionals

$$X = \begin{vmatrix} x^3 & x^1 - ix^2 \\ x^1 + ix^2 & -x^3 \end{vmatrix} \text{ и } Y = \begin{vmatrix} y^3 & y^1 - iy^2 \\ y^1 + iy^2 & -y^3 \end{vmatrix},$$

it is required to define:

1) a set of unitary matrices $C = \begin{vmatrix} \overline{\alpha} & -\beta \\ \overline{\beta} & \alpha \end{vmatrix}$ that satisfy the equality

$$Y = \overline{C}^T X C \quad (4)$$

2) one-dimensional subspaces that are invariant with respect to transformations represented by matrices C (i.e. a set of respective rotation centers)
   Note that since the transformation C is unitary, the vector norms defined by the determinants of matrices of the Hermitian functionals X and Y coincide, and, therefore, (4) defines rotation.

From equality (4) we can obtain the following system of linear homogeneous equations with respect to the unknown variables α and β:

$$\begin{aligned} x_3\alpha + (x_1 + ix_2)\beta &= y_3\alpha - (y_1 - iy_2)\overline{\beta} \\ (x_1 - ix_2)\alpha - x_3\beta &= y_3\beta + (y_1 + iy_2)\overline{\alpha} \end{aligned}. \quad (5)$$

For arbitrary α, a solution of (5) is given by

$$\beta = \frac{(x_1 - ix_2)\alpha - (y_1 - iy_2)\overline{\alpha}}{x_3 + y_3}. \quad (6)$$

From (5) we have

$$\operatorname{Re}\beta = \beta_1 = \frac{\alpha_1(x_1 - y_1) + \alpha_2(x_2 + y_2)}{x_3 + y_3}$$

and

$$\operatorname{Im}\beta = \beta_2 = \frac{\alpha_2(x_1 + y_1) - \alpha_1(x_2 - y_2)}{x_3 + y_3}. \quad (7)$$

Using one of the properties of unitarity of the matrix C ($\det C = \alpha_1^2 + \alpha_2^2 + \beta_1^2 + \beta_2^2 = 1$), we can define either $\alpha_1 = \operatorname{Re}\alpha$ or $\alpha_2 = \operatorname{Im}\alpha$. Note that one of these parameters remains arbitrary. Thus (6) defines rotation for $\alpha \neq 0$ and $x_3 + y_3 \neq 0$.

The invariance of the rotation center $z(z_1, z_2, z_3)$ with respect to the transformation C is written as a condition:

$$\overline{C}^T Z C = Z,$$

from which we obtain

$\beta_1 z_1 - \beta_2 z_2 = 0;\ \alpha_2 z_2 - \beta_1 z_3 = 0;\ \alpha_1 z_1 - \beta_2 z_3 = 0.$ It is not difficult to verify that the determinant of this system considered for the unknown values z_1, z_2 and z_3 is identically zero, and, therefore, for given α_1, α_2, β_1 and β_2 ($\alpha \neq 0$ and $x_3 + y_3 \neq 0$) there always exist nontrivial solutions written in the form

$$z_1 = \frac{\beta_2}{\alpha_2} z_3;\ z_2 = \frac{\beta_1}{\alpha_2} z_3, \quad (8)$$

where z_3 is arbitrary.

Thus, (7) together with the normalization property define a generalized rotation transforming (x^1,x^2,x^3) to (y^1,y^2,y^3) with respect to the set of centers which is defined by (8).

2.2. Relations between Transformations in C^2 and L^3

We can establish the correspondence between the elements of the transformation matrix $C = \begin{vmatrix} \overline{\alpha} & -\beta \\ \overline{\beta} & \alpha \end{vmatrix}$ in C^2 and the elements of the orthogonal real matrix of rotation A in L^3.

The matrix A is, by definition, the matrix of transformation between two orthonormalized basis vectors of the space L^3 , and its rows are decompositions of the new basis vectors in terms of the initial basis vectors. Hence, due to the identification of the spaces $L(C^2)$ and L^3 we have

$$\overline{C}^T \sigma_i C = a_i^{i'} \sigma_{i'} \quad (i, i' = 1,2,3), \tag{9}$$

where σ_i are the Pauli matrices corresponding to the initial basis, $\sigma_{i'}$ are the Pauli matrices of the new basis, and $\alpha_i^{i'}$ are the elements of the matrix A^{-1}.

Formula (10) can be written explicitly in the form of following three equalities

$$a_1^1\begin{vmatrix}0 & 1\\ 1 & 0\end{vmatrix} + a_2^1\begin{vmatrix}0 & -i\\ i & 0\end{vmatrix} + a_3^1\begin{vmatrix}1 & 0\\ 0 & -1\end{vmatrix} = \begin{vmatrix}\overline{\alpha} & -\beta\\ \overline{\beta} & \alpha\end{vmatrix} * \begin{vmatrix}0 & 1\\ 1 & 0\end{vmatrix} * \begin{vmatrix}\alpha & \beta\\ -\overline{\beta} & \overline{\alpha}\end{vmatrix},$$

$$a_1^2\begin{vmatrix}0 & 1\\ 1 & 0\end{vmatrix} + a_2^2\begin{vmatrix}0 & -i\\ i & 0\end{vmatrix} + a_3^2\begin{vmatrix}1 & 0\\ 0 & -1\end{vmatrix} = \begin{vmatrix}\overline{\alpha} & -\beta\\ \overline{\beta} & \alpha\end{vmatrix} * \begin{vmatrix}0 & -i\\ i & 0\end{vmatrix} * \begin{vmatrix}\alpha & \beta\\ -\overline{\beta} & \overline{\alpha}\end{vmatrix}.$$

$$a_1^3\begin{vmatrix}0 & 1\\ 1 & 0\end{vmatrix} + a_2^3\begin{vmatrix}0 & -i\\ i & 0\end{vmatrix} + a_3^3\begin{vmatrix}1 & 0\\ 0 & -1\end{vmatrix} = \begin{vmatrix}\overline{\alpha} & -\beta\\ \overline{\beta} & \alpha\end{vmatrix} * \begin{vmatrix}1 & 0\\ 0 & -1\end{vmatrix} * \begin{vmatrix}\alpha & \beta\\ -\overline{\beta} & \overline{\alpha}\end{vmatrix},$$

which readily yield the following expressions for calculating the elements of the matrix A by the elements of the matrix C:

$$a_1^1 = (\alpha_1^2 - \alpha_2^2) - (\beta_1^2 - \beta_2^2);\ a_2^1 = 2(\alpha_1\alpha_2 + \beta_1\beta_2);\ a_3^1 = 2(\alpha_2\beta_2 - \alpha_1\beta_1);$$
$$a_1^2 = 2(\beta_1\beta_2 - \alpha_1\alpha_2);\ a_2^2 = (\alpha_1^2 - \alpha_2^2) + (\beta_1^2 - \beta_2^2);\ a_3^2 = 2(\alpha_1\beta_2 + \alpha_2\beta_1);$$
$$a_1^3 = 2(\alpha_1\beta_1 + \alpha_2\beta_2);\ a_3^2 = 2(\alpha_2\beta_1 - \alpha_1\beta_2);\ a_3^3 = (\alpha_1^2 + \alpha_2^2) - (\beta_1^2 + \beta_2^2). \tag{10}$$

Expressions (10) enable the calculation of the elements of the matrix A through the given coordinates of three points (initial, terminal and the center) that define rotation. On the other hand, taking into account that the matrix A of basic representation can be written in the form [1]

$$A = \begin{Vmatrix}\cos\varphi\cos\psi - \cos\theta\sin\varphi\sin\psi & -\cos\varphi\sin\psi - \cos\theta\sin\varphi\sin\psi & \sin\varphi\sin\theta\\ \sin\varphi\cos\psi + \cos\theta\cos\varphi\sin\psi & -\sin\varphi\sin\psi + \cos\theta\cos\varphi\cos\psi & -\cos\varphi\sin\theta\\ \sin\psi\sin\theta & \cos\psi\sin\theta & \cos\theta\end{Vmatrix},$$

where $-\pi<\varphi\leq\pi$, $0\leq\theta\leq\pi$ и $-\pi<\psi\leq\pi$ are Euler angles, it easily follows that expressions (10) enable definition of Euler angles as well,

$$\cos\theta = a_{33};\ \sin\varphi\sin\theta = a_{31} \text{ and } \sin\psi\sin\theta = a_{13} \tag{11}$$

2.3. Control of Spatial Rotations

Having expressions (7) and (11), it is easy to calculate the Euler angles that ensure rotation of the point $x(x^1, x^2, x^3)$ to the point $y(y^1, y^2, y^3)$. If it is assumed that to the initial point $x(x^1, x^2, x^3)$ there correspond the zero Euler angles $\theta_0 = \phi_0 = \psi_0 = 0$, then the control of rotation consists in making a time-dependent change of the Euler angles from the initial values $\theta_0; \phi_0; \psi_0$ to the terminal values $\theta_f; \phi_f; \psi_f$ calculated by formulas (11).

In a general form, the control process can be represented as change functions of the Euler angles $\theta(t)$; $\phi(t)$; $\psi(t)$ which must satisfy the conditions

$$\theta(t_0) = 0;\ \phi(t_0) = 0;\ \psi(t_0) = 0,$$
$$\theta(t_f) = \theta_f;\ \phi(t_f) = \phi_f;\ \psi(t_f) = \psi_f,$$

where t_0 and t_f are the initial and terminal moments of time.

The above-said naturally implies the problem of defining the control functions $\theta(t)$; $\phi(t)$; $\psi(t)$.

It should be emphasized that dependences $\theta(t)$; $\phi(t)$; $\psi(t)$ have a kinematic character, since they take into account neither moments, nor elasticities, nor any other dynamic characteristics of the process, and therefore, after defining them, there arises a problem of synthesizing – on the basis of these functions – the dynamic adaptive control. This issue is discussed in [4].

With initial and final state vectors $x(x^1, x^2, x^3)$ and $y(y^1, y^2, y^3)$ corresponding, there is also the intermediate rotating vector $\xi(\xi^1, \xi^2, \xi^3)$, which at the initial moment of time $t = t_0$ coincides with the initial rotation vector $x(x^1, x^2, x^3)$ and, at the terminal moment of time $t = t_f$, with the terminal vector $y(y^1, y^2, y^3)$. The moving angle γ between the vectors $x(x^1, x^2, x^3)$ and $\xi(\xi^1, \xi^2, \xi^3)$ is equal, at the initial moment of time $t = t_0$, to zero and, at the moment of time $t = t_f$, to $\gamma = \gamma_f$, where

$$\gamma_f = ar\cos\left(\frac{(x, y)}{|x| * |y|}\right) = ar\cos\left(\frac{(x, y)}{|x|^2}\right)^1$$

and (x, y) is the dot product of the vectors x and y.

It is obvious that the moving angle between the vectors $y(y^1, y^2, y^3)$ and $\xi(\xi^1, \xi^2, \xi^3)$ is equal to $\gamma_f - \gamma$.

---

[1] $|x| * |y| = |x|^2$, since during rotation $|x| = |y|$.

Let us define the coordinates of the vector $\xi(\xi^1,\xi^2,\xi^3)$ assuming that it forms the angles γ and $\gamma_f-\gamma$ with the vectors $x(x^1,x^2,x^3)$ and $y(y^1,y^2,y^3)$ and is located in their plane. To this end, we introduce the vector $r=(x^2y^3-x^3y^2,\ x^3y^1-x^1y^3,\ x^1y^2-x^2y^1)$, which is the cross product of the vectors x and y. Then the above conditions can be written in the form of the following system of linear equations:

$$
\begin{aligned}
(\xi,r)&=0;\\
(\xi,x)&=|x|^2\cos\gamma;\\
(\xi,y)&=|x|^2\cos(\gamma_f-\gamma).
\end{aligned}
\tag{12}
$$

It is not difficult to see that the vector $\xi(\xi^1,\xi^2,\xi^3)$ defined from system (12) satisfies the following conditions:

1. For $\gamma=0$, $\xi(\xi^1,\xi^2,\xi^3)=x(x^1,x^2,x^3)$, which follows from the second equation of system (12), since in this case $(\xi,x)=|x|^2$, which is possible only provided that $\xi(\xi^1,\xi^2,\xi^3)=x(x^1,x^2,x^3)$;
2. For $\gamma=\gamma_f$, $\xi(\xi^1,\xi^2,\xi^3)=y(y^1,y^2,y^3)$, which follows from the third condition of system (12), since in this case $(\xi,y)=|x|^2$, which is possible only provided that $\xi(\xi^1,\xi^2,\xi^3)=y(y^1,y^2,y^3)$;
3. $|\xi|=|x|=|y|$, which follows from the second and third equations of system (12).

Therefore the vector $\xi(\xi^1,\xi^2,\xi^3)$ defined from system (13) corresponds to Fig. 1, i.e., it can actually be regarded as the vector rotating (condition 3) from the vector $x(x^1,x^2,x^3)$ (condition 1) to the vector $y(y^1,y^2,y^3)$ (condition 2). Note that in this case the angle γ changes in within $0\le\gamma\le\gamma_f$.

The equations of system (12) can be written in the coordinate form as follows:

$$
\begin{aligned}
\xi^1r^1+\xi^2r^2+\xi^2r^2&=0\\
\xi^1x^1+\xi^2x^2+\xi^2x^2&=|x|^2\cos\gamma\\
\xi^1y^1+\xi^2y^2+\xi^2y^2&=|x|^2\cos(\gamma_f-\gamma).
\end{aligned}
$$

It is not difficult to see that its determinant is equal to

$$
\Delta=\begin{vmatrix} r^1 & r^2 & r^3\\ x^1 & x^2 & x^3\\ y^1 & y^2 & y^3\end{vmatrix}=|r|^2.
$$

Other determinants of Kramer's formulas for system (12) will be equal to

$$\Delta_1 = \begin{vmatrix} 0 & r^2 & r^3 \\ |x|^2 \cos\gamma & x^2 & x^3 \\ |x|^2 \cos(\gamma^f - \gamma) & y^2 & y^3 \end{vmatrix} = \qquad ;$$

$$|x|^2((\cos(\gamma_f - \gamma)(r^2x^3 - r^3x^2) - \cos\gamma(r^2y^3 - r^3y^2))$$

$$\Delta_2 = \begin{vmatrix} r^1 & 0 & r^3 \\ x^1 & |x|^2 \cos\gamma & x^3 \\ y^1 & |x|^2 \cos(\gamma_f - \gamma) & y^3 \end{vmatrix} = \qquad ;$$

$$|x|^2((\cos(\gamma_f - \gamma)(r^3x^1 - r^1x^3) - \cos\gamma(r^3y^1 - r^1y^3))$$

$$\Delta_3 = \begin{vmatrix} r^1 & r^2 & 0 \\ x^1 & x^2 & |x|^2 \cos\gamma \\ y^1 & y^2 & |x|^2 \cos(\gamma_f - \gamma) \end{vmatrix} = \qquad .$$

$$|x|^2((\cos(\gamma_f - \gamma)(r^1x^2 - r^2x^1) - \cos\gamma(r^1y^2 - r^2y^1))$$

If we introduce the new vectors $r_x = (r^2x^3 - r^3x^2; r^3x^1 - r^1x^3; r^1x^2 - r^2x^1)$ and $r_y = (r^2y^3 - r^3y^2; r^3y^1 - r^1y^3; r^1y^2 - r^2y^1)$, which equal to the vector products [r×x] and [r×y], respectively, then the coordinates of the rotating vector will be presented in the following form:

$$\xi^i(t) = \frac{|x|^2}{|r|^2}(\cos(\gamma_f - \omega t)r_x^i - \cos\omega t r_y^i) \quad (i = 1,2,3) \tag{13}$$

In these expressions, the angle γ is an independent variable and can be treated as time function, which means that the coordinates of the vector $\xi(\xi^1, \xi^2, \xi^3)$ are also time functions. We would like to emphasize that the problem of synthesis of spatial motion control thus reduces to defining a function $\gamma(t)$ of the concrete form, which is connected with the rotation process dynamics and was discussed in [4].

Here we assume that $\gamma(t)$ is sufficiently smooth and satisfies the conditions $\gamma(t = t_o) = 0$ *u* $\gamma(t = t_f) = \gamma_f$.

For definiteness, we assume that $\gamma(t) = \omega t$, where $\omega = 2\pi f$ is the constant angular velocity.

As has already been noted, the vector $\xi(\xi^1, \xi^2, \xi^3)$ is a rotating vector, and, therefore, at each moment of time it can be considered as a terminal vector of the current moment of the rotation process. If in formulas (7) we replace the coordinates of the point $y(y^1, y^2, y^3)$ by expressions (14), then we obtain representations of the parameters of the spinor matrix *C*, the orthogonal matrix *A* (formulas (11)) and Euler angles (12) in terms of time functions. Thus, we obtain a time-dependent (kinematics) representation of the rotation of the point $x(x^1, x^2, x^3)$ to the point $y(y^1, y^2, y^3)$. However, first, we should predetermine the matrix C so that at the initial moment of time the spinor equation of rotation (4) would have the form $X = \overline{C}^T XC$, which is evidently possible only if C is a unit matrix. This can be done by an appropriate choice of the parameters α_1 and α_2.

Indeed, setting $\alpha_1 = 1$; $\alpha_2 = 0$ we obtain

$$\operatorname{Re}\beta = \beta_1 = \frac{x^1 - y^1}{x^3 + y^3}; \ \operatorname{Im}\beta = \beta_2 = \frac{y^2 - x^2}{x^3 + y^3}.$$

By substituting the last formulas and (14) for corresponding elements of the matrix C it will take the following form:

$$C(t) = \frac{1}{1+|\beta|^2}\begin{vmatrix} 1 & \dfrac{((\xi^1(t)-x^1)+i(\xi^2(t)-x^2))}{x^3+\xi^3(t)} \\ \dfrac{((x^1-\xi^1(t))-i(x^2-\xi^2(t)))}{x^3+\xi^3(t)} & 1 \end{vmatrix}, \tag{14}$$

where $|\beta|^2 = \dfrac{(x^1 - \xi^1(t))^2 + (x^2 - \xi^2(t))^2}{(x^3 + \xi^3(t))^2}$.

Note that detC(t)=1 for any t.

It is obvious that at the initial moment of time t_0 the matrix $C(t = t_o) = \begin{vmatrix} 1 & 0 \\ 0 & 1 \end{vmatrix}$, since in that case $\gamma(t_0) = 0$ and $\xi(\xi^1; \xi^2; \xi^3) = x(x^1, x^2, x^3)$.

For $t = t_f$ we have $\gamma(t_f) = \gamma_f$, $\xi(\xi^1; \xi^2; \xi^3) = y(y^1, y^2, y^3)$.

From the above-said, it follows that the obtained spinor matrix of rotation (14) is defined correctly. However, in that case the Euler angles (11), too, are defined correctly. They also turn out to be the functions of time

$$\theta(t) = \arccos\left(\frac{(x^3 + \xi^3(t))^2 - (x^1 - \xi^1(t))^2 - (x^2 - \xi^2(t))^2}{(x^3 + \xi^3(t))^2}\right);$$

$$\varphi(t) = \arcsin\left(\frac{2(x^1 - \xi^1(t))}{(x^3 + \xi^3(t))\sin\theta(t)}\right):$$

$$\psi(t) = \arcsin\left(\frac{2(\xi^1(t) - x^1)}{(x^3 + \xi^3(t))\sin\theta(t)}\right). \tag{15}$$

Expressions (15) solve the problem we have formulated on defining the kinematics functions $\theta(t)$; $\phi(t)$; $\psi(t)$. On the other hand, it should be noted that the proposed theory allows one to reduce an actually three-dimensional problem of spatial motion control to a one-dimensional problem. Indeed, for this it is sufficient to synthesize in one way or another function $\gamma(t)$ satisfying the corresponding boundary conditions [5]. Then it is obvious that the control process is completely defined by the spinor matrix of rotation (14) and the Euler angle functions (15).

Conclusion

Let us consider a numerical example illustrating and concluding the above reasoning. Assume that the initial vector $x(10, -45, 30)$ and the terminal vector $y(1, 20, 51.225)$ are given arbitrarily. The angle between them is equal to $\gamma_f = ar\cos\left(\frac{(x,y)}{|x|^2}\right) = 77.65^0$.

Assuming for the sake of simplicity that $\omega = 1$ with a step equal to $\frac{\gamma_f}{3}$, lets us calculate in two different ways three intermediate positions of the rotating vector $\xi(\xi^1;\xi^2;\xi^3)$. Using formulas (14), we obtain the following coordinates of the rotating vector for three angle values: (8.48, -27.25, and 47.02) - angle 25.88^0; (5.27, -4.03, and 54.50) - angle 51.77^0; and (1.00, 20.00, 51.23) - angle 77.65^0.

The procedure of verifying whether the Euler angles have been calculated consists of the following: using the obtained coordinates of the intermediate positions of the vector $\xi(\xi^1;\xi^2;\xi^3)$, for each of three angle values given above we should calculate the Euler angles by formulas (16) and the three-dimensional orthogonal matrix A of the basic representation and then again the intermediate coordinates of the rotating vector by the formula $\xi = Ax$, where x is the initial rotation vector. The obtained values should coincide in both cases. The results of the corresponding calculations are presented in Table 1.

Table 1.

Angle γ	**Coordinates of rotating vector (formulas (14))**			**Norm of ξ**
	ξ^1	ξ^2	ξ^3	$\lvert\xi\rvert$
$\frac{\gamma_f}{3}$ =25.88^0	8.48	–27.25	47.02	55
$\frac{2\gamma_f}{3}$ =51.77^0	5.2	–4.03	54.60	55
γ_f=77.65^0	1	20	51.23	55

The matrix A was calculated for the Euler three angle values by formulas (15).

The coordinate values of the rotating vector $\xi(\xi^1;\xi^2;\xi^3)$ were calculated by multiplying matrix A by the initial rotation vector $x(10, -45,30)$: $\xi = Ax$.

From Table 1 we see that the coordinates of the rotating vector coincide with the coordinates calculated by formulas (13).

References

[1] K. S. Fu., R.C. Gonzalez, and K, Lie. *Robotics: Control Sensing, Vision and Intelligence*, McGraw-Hill Book Company, 1987.

[2] Y. Tokunaga, T. Hakukawa, T. Inoue, "Algorithm and Design of an Robot", *J. of Robotics and Mechatronics*, Vol.11, No.3, 1999, pp.173-182.

[3] Milnikov A.A., Prangishvili A.I., Rodonaia I.D. *A Spinor Model of 3-Dimensional Generalized Rotations, Automation and Remote Contr*ol, 2005, № 6, p. 30-37.

[4] Milnikov A.A., New Procedure of Robots Spatial Rotations Terminal Control. *13th International Conference on Applied Mechanics and Mechanical Engineering.* Cairo, Egypt, 2008.

[5] I.M. Gelfund, R.A. Minlos, and Z.Ya. Shapiro, *Representations of the Group of Rotations and the Lorenz Group*. Fizmatgiz, Moscow, 1958 (in Russian).

In: Several Problems of Applied Mathematics and Mechanics ISBN: 978-1-62081-603-5
Editors: Ivane Gorgidze and Tamar Lominadze

Chapter 4

ON THE RELATIONSHIP BETWEEN QUALITATIVE AND QUANTITATIVE IMPORTANCE OF CRITERIA IN MULTICRITERIAL DECISION MAKING PROBLEMS

***A. P. Nelyubin*[1] *and V. V. Podinovski*[2]**
[1]Institute of Machines Science of the RAS, Moscow, Russia
[2]National Research University "Higher School of Economics", DeCAn Lab, Moscow, Russia

Abstract

The influence of the assumption about the existence of cardinal coefficients of criteria importance, consistent with the importance ordering of criteria (according to the definitions in the criteria importance theory), on the preference relation generated by this information, is investigated.

Keywords: multiple criteria decision making problems, criteria importance theory, ordering of criteria according to importance, coefficients of importance

1. Introduction

The concept of the relative importance, or significance, of criteria is fundamental to most of the well-known methods of multiple criteria decision analysis. However, no rigorous definition of this concept has been introduced by the authors of such methods, who instead relied on an intuitive understanding of the concept [1, 2].

The mathematical theory of criteria importance [2 – 9] relies on strict definitions of the concepts of qualitative importance: 'one criterion is more important than the other', 'both criteria are equally important'; as well as quantitative importance: 'one criterion is so many times more important than the other'. In this theory, decision rules were developed that induce the preference relations on the basis of the qualitative or quantitative information about the importance of criteria, taking into account the information about the change of preferences along the criteria scale. However until recently, there was an open question whether the preference relation induced by the qualitative importance information extends at the expense of the assumption about the existence of the quantitative importance coefficients or not. This paper is devoted to answering this question representing both the theoretical and practical interest.

2. The Basic Mathematical Model

The following presentation is based on the given mathematical model:

$$\mathcal{M} = < \tau, X, f, Z, R >,$$

where τ is the decision problem formulation (choose one or several best variants, sort all variants by preference etc.), X is the set of variants (in specific problems, these could be strategies, variants, outcomes, consequences), $f = (f_1, \ldots, f_m)$ is the vector criterion, f_i are individual criteria ($m \geq 2$), Z is the set of estimates of the vector criterion, and R is a non-strict preference relation of the decision maker (DM). Every variant x from *the set of all variants X* is characterized by its values on criteria f_i. *Criterion* f_i is a function defined on X and takes its values from set Z_i. This set is referred to as the scale, set of estimates, grades, or values of criterion f_i. Below, it is assumed that all the criteria are homogeneous, i.e. have the same scale (which can be an ordinal scale) and, particularly, the same set of values $Z_0 = \{1, \ldots, q\}$, where $q \geq 2$. Thus, every variant x is characterized by the values $f_i(x)$ of all the criteria that together form the *vector estimate* $y = f(x) = (f_1(x), \ldots, f_m(x))$ of this variant. Therefore, the comparison of variants by preference is reduced to the comparison of their vector estimates. The set of all vector estimates is $Z = Z_0^m$.

The preferences of the DM are modeled using *the non-strict preference relation R* on set Z: yRz means that vector estimate y is not less preferable than z. Relation R induces the (strict) *preference relation P* and *indifference I*:

$$yIz \Leftrightarrow yRz \wedge zRy; \; yPz \Leftrightarrow yRz \wedge z\not{R}y$$

($z\not{R}y$ means that zRy is not true). The binary relation R is a (partial) quasi-order, i.e. it is reflexive and transitive: relation yRy holds for any vector estimate y, and yRz and zRu imply yRu.

Relation R is constructed using the information about the preferences of the DM. It is assumed that preferences increase on the scale Z_0 so that the (unknown) values $v_0(k)$ of gradations k satisfy the inequalities:

$$v_0(1) < v_0(2) < \ldots < v_0(q), \tag{2.1}$$

i.e. the value function v_0 is increasing on Z_0. If there are no other restrictions on values of this function, then the scale of criteria is ordinal. Taking into account (2.1) the Pareto relation, R^0 is defined on the set of vector estimates Z:

$$yR^0z \Leftrightarrow (y_i \urcorner z_i,\ i = 1, \ldots, m).$$

Preferences of the DM are consistent with the Pareto relation: $P^0 \subset P$.

To solve the multiple criteria decision problem under given formulation only by means of the Pareto relation, as a rule, is not possible, therefore it is required to "expand" it taking into account additional information about the preferences of the DM. Below, we consider the case in which such information about the relative importance of criteria and the criteria scale is given.

3. Some Facts from the Criteria Importance Theory

For convenience, below we present basic definitions and some results from the criteria importance theory, first from the theory of qualitative importance [2 – 6]. Denote y^{ij}, the vector obtained from vector $y = (y_1, \ldots, y_m)$, by permuting its components y_i and y_j. For example, if $y = (2, 3, 5, 4)$, then $y^{13} = (5, 3, 2, 4)$.

Definition 1. The statement "criteria f_i and f_j are equally important", which is denoted as $i \approx j$, means that any two vector estimates y and y^{ij} are indifferent.

Definition 2. The statement "criterion f_i is more important than criterion f_j", which is denoted as $i \succ j$, means that any vector estimate y, such that $y_i > y_j$, is preferred to y^{ij}.

Let Ω be the qualitative importance information, i.e. the set of all available statements of the form $i \approx j$ and $i \succ j$. According to definitions 1 and 2, the statement $i \approx j$ induces the indifference relation $I^{i \approx j}$ on the set of vector estimates Z, and the statement $i \succ j$ induces the preference relation $P^{i \succ j}$ on the set Z, which are defined as follows:

$$yI^{i \approx j}z \Leftrightarrow (z = y^{ij},\ y_i \neq y_j),\ yP^{i \succ j}z \Leftrightarrow (z = y^{ij},\ y_i > y_j).$$

The non-strict preference relation (quasi-order) R^{Ω} induced on Z by qualitative information about criteria preference Ω, is defined as the least transitive relation which contains the Pareto relation R^0 and relations R^{ω} for all $\omega \in \Omega$ (here $R^{\omega} = I^{i \approx j}$ if $\omega = i \approx j$, and $R^{\omega} = I^{i \succ j}$ if $\omega = i \succ j$):

$$R^{\Omega} = \mathrm{TrCl}[(\textstyle\bigcup_{\omega \in \Omega} R^{\omega}) \bigcup R^0],$$

where TrCl denotes the operation of transitive closure of a binary relation. According to this definition relation, $yR^{\Omega}z$ holds if and only if there exists a sequence such that

$$y\, R^{\omega^1}\, u^1, u^1\, R^{\omega^2}\, u^2, \ldots, u^{r\text{-}1}\, R^{\omega^r}\, z; \tag{3.1}$$

here, u^k – vector estimates and ω^k are either information ω from Ω or the symbol 0.

Definition 2. The numbers $\beta_1, \ldots, \beta_m$ are called *importance index* of criteria, consistent with the information Ω, if they satisfy the conditions:

$$i \approx j \in \Omega \Rightarrow \beta_i = \beta_j,\ i \succ j \in \Omega \Rightarrow \beta_i > \beta_j.$$

If the importance indexes are positive and in total are equal to 1, then they are called *coefficients of importance* of criteria and denoted as α_i.

Information Ω is *complete* and *consistent* if it is sufficient for a complete ordering of all criteria according to their importance. Such information will be denoted as Ω^*. The importance indexes and coefficients of importance of criteria, consistent with the information Ω^*, are called *ordinal*.

Below, we will use the following notations for vectors from Re^n ($n > 1$):

$$a \geqq b \Leftrightarrow a_i \geq b_i,\ i = 1,\ldots, n;\ a \geq b \Leftrightarrow a \geqq b,\ a \neq b;\ a > b \Leftrightarrow a_i > b_i,\ i = 1,\ldots, n;$$

$a_{\uparrow} = (a_{(1)}, \ldots, a_{(n)})$ is the vector obtained by permuting the components of vector a in the non-increasing order, i.e. $a_{(1)} \geq \ldots \geq a_{(n)}$.

For information Ω^*, the relation R^{Ω^*} can be specified analytically.

$$\alpha_{ik}(y) = \begin{cases} \alpha_i, & y_i = k, \\ 0, & y_i \neq k, \end{cases} \quad \sigma_i^k(y) = \sum\nolimits_{j=1}^{k} \alpha_{ij}(y),\ i = 1, \ldots, m,\ k = 1, \ldots, q.$$

The following statement holds:

$$yR^{\Omega^*}z \Leftrightarrow \sigma_{\uparrow}^k(y) \leqq \sigma_{\uparrow}^k(z),\ \ k = 1,\ldots, q-1. \tag{3.2}$$

If all the non-strict inequalities $\leqq$ are satisfied as equalities in (3.2), then $yI^{\Omega^*}z$, and if at least one of the $\leqq$ is $<$, then $yP^{\Omega^*}z$.

Assume that $q > 2$ and there is an additional information D that slows down the rate of growth of preferences (so criteria have the first ordered metric scale [10]):

$$v_0(2) - v_0(1) > v_0(3) - v_0(2) > \ldots > v_0(q) - v_0(q-1). \tag{3.3}$$

At the presence of information D, statement $i \approx j$ also induces the preference relation $P^{i \approx j \& D}$ on the set of vector estimates Z, and statement $i \succ j$ induces the preference relation $P^{i \succ j \& D}$ on the set Z, which are defined as follows:

$$\begin{aligned} yP^{i \approx j \& D}z \Leftrightarrow & (y = (z \,\|\, z_i + l, z_j - l),\ z_i + l \geq z_j - l) \vee \\ & (y = (z \,\|\, z_j + l, z_i - l),\ z_j + l \geq z_i - l), \end{aligned} \tag{3.4}$$

$$yP^{i \succ j \& D}z \Leftrightarrow (y = (z \,\|\, z_i + l, z_j - l),\ z_i + l \geq z_j - l), \tag{3.5}$$

where $l \in Z_0$, $(z \| z_i + l, z_j - l)$ is a vector estimate, obtained from z by increasing the component z_i to $z_i + l$ and decreasing the component z_j to $z_j - l$. The sense of these relations is simple. Definition (3.4) states: if criteria f_i and f_j are equally important, then by increasing the lesser from components z_i and z_j by a natural number l and simultaneously decreasing the greater by the same number l, such that the increased component does not become greater than the decreased one, we will obtain a vector estimate which is more preferable than the initial vector estimate z. This definition is schematically illustrated in Figure 1.

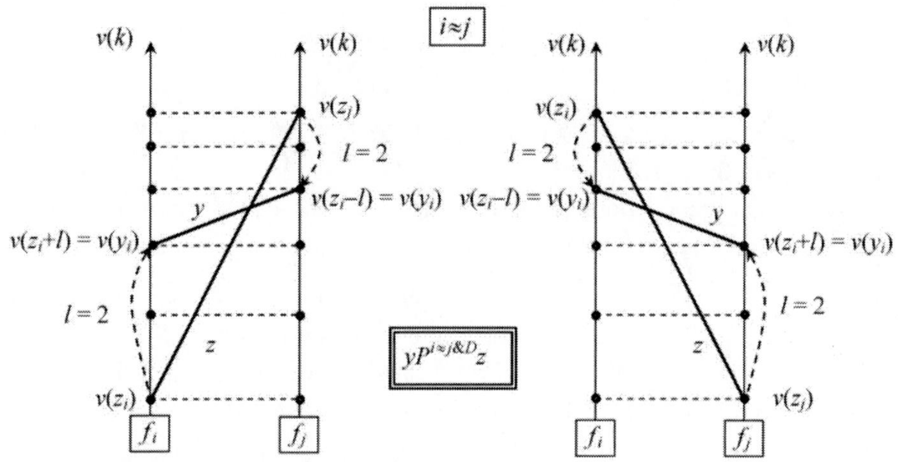

Figure 1. Schematic illustration to the definition of the relation $P^{i \approx j \& D}$.

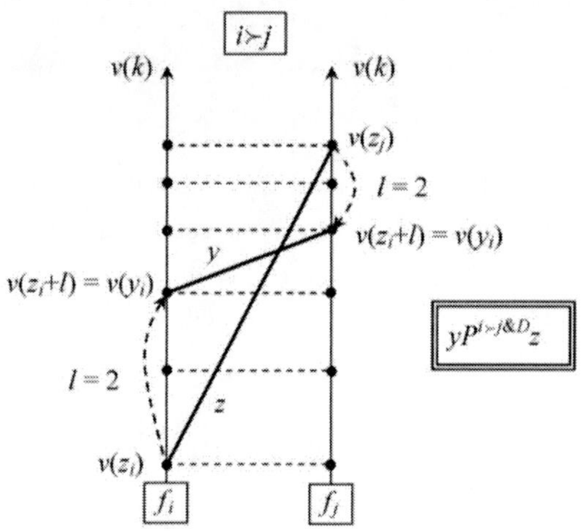

Figure 2. Schematic illustration to the definition of the relation $P^{i \succ j \& D}$.

Definition (3.4) states: if some criterion f_i is more important than f_j and the component z_i of a vector estimate z is less than the component z_j, then by increasing z_i by a natural number l and simultaneously decreasing z_j by the same number l, so that the increased component does not become greater than the decreased one, we will obtain a vector estimate which is more preferable than the initial vector estimate z. This definition is schematically illustrated in Figure 2.

In the scope of the problem under the study definitions, (3.4) and (3.5) can also be considered as definitions of the first ordered metric scale with a slowdown of the rate of growth of preferences along the criteria scale without using inequalities (3.3).

The non-strict preference relation (quasi-order) $R^{\Omega\&D}$ induced on Z by the qualitative information about the criteria preference Ω and the information D is defined as follows:

$$R^{\Omega\&D} = \mathrm{TrCl}[(\bigcup_{\omega\in\Omega}(R^{\omega} \cup P^{\omega\&D}) \cup R^{0}],$$

where $R^{\omega} = P^{i\succ j}$ and $P^{\omega\&D} = P^{i\succ j\&D}$ if ω is $i\succ j$; $R^{\omega} = I^{i\approx j}$ and $P^{\omega\&D} = P^{i\approx j\&D}$ if ω is $i\approx j$. According to this definition, $yR^{\Omega\&D}z$ holds if and only if there exists a sequence (3.1), in which u^k are vector estimates, and ω^k is ω or $\omega\&D$ (so that R^{ω^k} is either I^{ω^k} or P^{ω^k} subject to the sense of ω^k) or 0.

Now we present some information from the theory of quantitative importance [8, 9]. Quantitative information about criteria importance Θ contains statements like "Criterion f_i is h_{ij} times as important as criterion f_j" (this statement is denoted as $i \succ^{h_{ij}} j$) or estimates (e.g. interval) of values h_{ij}. We do not formulate the exact definition of the concept that one criterion is h times as important as the other, because we will not use it below. Quantitative information about criteria importance is called *complete and consistent* and denoted as Θ^*, if it consists of statements $i \succ^{h_{ij}} j$ and induces *quantitative*, or *cardinal indexes of criteria importance* β_i – positive numbers satisfying the conditions:

$$h_{ij} = \frac{\beta_i}{\beta_j}, \; i, j = 1, \dots, m.$$

If the total of all β_i is equal to 1, then they are called *quantitative*, or *cardinal coefficients of importance* of criteria and denoted as α_i. These coefficients are determined explicitly by the information Θ^*, i.e. the vector $\alpha = (\alpha_1, \dots, \alpha_m)$ is single.

The decision rule which defines the non-strict relation R^{Θ^*} induced by information Θ^* can be stated in the following form:

$$yR^{\Theta^*}z \Leftrightarrow \sum_{i=1}^{m} \sigma_i^k(y) \geq \sum_{i=1}^{m} \sigma_i^k(z) \;, k = 1, \dots, q-1. \quad (3.6)$$

If all the non-strict inequalities are satisfied as equalities in (3.6), then $yI^{\Theta^*}z$, and if at least one of the inequalities is strict, then $yP^{\Theta^*}z$.

If information Θ is incomplete and the exact values of criteria importance coefficients α_i are unknown (e.g. there are only interval estimates of these values), then we have only a (nonempty) set A of possible values of vector $\alpha = (\alpha_1, \dots, \alpha_m)$. The relation R^{A} corresponding to such an information is defined as: relation $yR^{\mathrm{A}}z$ holds if and only if all of $q - 1$ inequalities in (3.6) are satisfied for any $\alpha \in \mathrm{A}$, or the equivalent:

$$yR^{\mathrm{A}}z \Leftrightarrow \sup_{\alpha\in \mathrm{A}} \sum_{i=1}^{m} (\sigma_i^k(y) - \sigma_i^k(z)) \geq 0, k = 1, \ldots, q-1, \tag{3.7}$$

Assume that there is an additional information D on which the rate of growth of preferences along the criteria scale slows down, so the inequalities (3.3) hold. The decision rule which defines the non-strict relation $R^{\Theta^*\&D}$ induced by information $\Theta^*\&D$ can be written as:

$$yR^{\Theta^*\&D}(\beta)z \Leftrightarrow \sum_{j=1}^{k}\sum_{i=1}^{m}\sigma_i^j(y) \geq \sum_{j=1}^{k}\sum_{i=1}^{m}\sigma_i^j(z) \;, \quad k = 1, \ldots, q-1. \tag{3.8}$$

If all the non-strict inequalities are satisfied as equalities in (3.8), then $yI^{\Theta^*\&D}$, and if at least one of the inequalities is strict, then $yP^{\Theta^*\&D}z$.

If the information Θ determines only a set A of possible values of α, then the relation $R^{\mathrm{A}\&D}$ is defined in the following form: relation $yR^{\mathrm{A}\&D}z$ holds if and only if all of $q-1$ inequalities in (3.8) are satisfied for any $\alpha\in\mathrm{A}$, or the equivalent:

$$yR^{\mathrm{A}\&D}z \Leftrightarrow \sup_{\alpha\in \mathrm{A}} \sum_{j=1}^{k}\sum_{i=1}^{m} (\sigma_i^j(y) - \sigma_i^j(z)) \geq 0, k = 1, \ldots, q-1. \tag{3.9}$$

4. Correlation between Qualitative and Quantitative Importance of Criteria

Information about qualitative criteria importance Ω^* and information about quantitative criteria importance Θ (in particular Θ^*) are consistent if:

$$i\approx j \Rightarrow \forall \alpha \in \mathrm{A} : \alpha_i = \alpha_j, \;, i\succ j \Rightarrow \forall \alpha \in \mathrm{A} : \alpha_i > \alpha_j;$$

$$(i\approx j \in \Omega^* \Rightarrow i \succ^1 j \,, i\succ j \in \Omega^* \Rightarrow i \succ^{h_{ij}} j, h_{ij} > 1).$$

Notice that for consistent Ω^* and Θ (in particular Θ^*) the following enclosures hold [2, 3, 8, 9]:

$$R^{\Omega^*} \subseteq R^{\mathrm{A}}, I^{\Omega^*} \subseteq I^{\mathrm{A}}, P^{\Omega^*} \subseteq P^{\mathrm{A}}; R^{\Omega^*\&D} \subseteq R^{\mathrm{A}\&D}, I^{\Omega^*\&D} \subseteq I^{\mathrm{A}\&D}, P^{\Omega^*\&D} \subseteq P^{\mathrm{A}\&D}. \tag{4.1}$$

$$(R^{\Omega^*} \subseteq R^{\Theta^*}, I^{\Omega^*} \subseteq I^{\Theta^*}, P^{\Omega^*} \subseteq P^{\Theta^*}; R^{\Omega^*\&D} \subseteq R^{\Theta^*\&D}, I^{\Omega^*\&D} \subseteq I^{\Theta^*\&D}, P^{\Omega^*\&D} \subseteq P^{\Theta^*\&D}).$$

Consider the case when the set $\mathrm{A} = \mathrm{A}^*$, in addition to constraints:

$$\alpha_i > 0, i = 1, \ldots, m; \alpha_1 + \ldots + \alpha_m = 1, \tag{4.2}$$

is determined by equalities and inequalities corresponding to the information Ω^* which sorts all the criteria by their importance:

$$i{\approx}j \in \Omega^* \Rightarrow \alpha_i = \alpha_j,\ i{\succ}j \in \Omega^* \Rightarrow \alpha_i > \alpha_j. \tag{4.3}$$

In essence putting the set A^* into consideration means that in addition to qualitative information about criteria importance Ω^*, we make an assumption about the existence of cardinal coefficients of criteria importance. Therefore, the question whether the non-strict enclosures (4.1) corresponding to A^* are satisfied as equalities represents both theoretical and practical interest. It was found that the answer to this question is positive for criteria with an ordinal scale, but negative for criteria with a first ordered metric scale!

Theorem 1. Equalities $R^{\Omega*} = R^{\mathrm{A}*}$, $I^{\Omega*} = I^{\mathrm{A}*}$, $P^{\Omega*} = P^{\mathrm{A}*}$ hold.

Theorem 2. Enclosures $R^{\Omega*\&D} \subseteq R^{\mathrm{A}*\&D}$, generally speaking, are strict.
The proofs of these theorems are given in the Appendix.

Conclusion

In this paper, we have shown that non-strict relations induced by qualitative information about criteria importance on making additional assumptions about the existence of quantitative coefficients of criteria importance does not expand in the case of ordinal scale of criteria and, generally speaking, expands in case of first ordered metric scale.

This result should be taken into account when using the known approach to a modeling of the preferences, based on usage of multiple estimations of importance coefficients values at the incomplete information about criteria importance and character of change of preferences along their scale.

Appendix

Proof of Theorem 1.

Without loss of generality, we will assume that according to the information Ω^*, all the criteria are arranged and indexed in descending order of their relative importance (the criterion f_1 is the most important one, and f_m is the least important).

Since the set A^* specified by (4.2) is not empty, instead of it in (3.7), we can use [11] the convex polyhedron set $\overline{\mathrm{A}}$, which is specified by the system:

$$\alpha_1 \rceil \alpha_2 \rceil \ldots \rceil \alpha_m \rceil 0;\ \alpha_1 + \ldots + \alpha_m = 1, \tag{6.1}$$

where $\alpha_i = \alpha_{i+1}$ if $i{\approx}i{+}1 \in \Omega^*$. After such a transition, we can use operator max instead of sup in the decision rule (3.7), which now can be written as follows:

$$yR^{\bar{A}}z \Leftrightarrow \max_{\alpha\in\bar{A}} \sum_{i=1}^{m}(\sigma_i^k(y)-\sigma_i^k(z)) \leq 0,\ k=1, \ldots, q-1. \tag{6.2}$$

Define the differences from (6.2) as functions of α:

$$\Delta_k(\alpha) = \sum_{i=1}^{m}(\sigma_i^k(y)-\sigma_i^k(z)) = \sum_{i=1}^{m}\alpha_i c_{ik}(y,z),$$
$$k=1, \ldots, q-1, \tag{6.3}$$

where numbers $c_{ik}(y,z) \in \{-1, 0, 1\}$ are determined by the vector estimates to compare.

Since the set $\bar{A}$ is a convex polyhedron and functions (6.3) are linear, problems (6.2) of their maximization are linear programming problems. Hence, the maximum of each of these functions on $\bar{A}$ is attained at one of the extreme points of this polyhedron.

The extreme points can be described analytically. If (6.1) does not contain any equality like $\alpha_i = \alpha_{i+1}$, then coordinates of an extreme point α^* are described as follows (see, e.g. [12]):

$$\alpha_1{}^* = \alpha_2{}^* = \ldots = \alpha_i{}^* = \frac{1}{i},\ \alpha_{i+1}{}^* = \ldots = \alpha_m{}^* = 0,\ i=1, \ldots, m. \tag{6.4}$$

Since the number of extreme points is equal to m, expression (6.4) describes a one-to-one correspondence between the numbers $i \in \{1, \ldots, m\}$ and the extreme points α^*, which can be denoted as $\alpha^*(i) = (\frac{1}{i}, \ldots, \frac{1}{i}, 0, \ldots, 0)$.

In cases when constraints (6.1) contain one equality ($\alpha_i = \alpha_{i+1}$), the number of extreme points is reduced by one. From two extreme points ($\alpha^*(i)$ and $\alpha^*(i+1)$), only the second remains. It follows from the fact that from all m vectors described in (6.4), only in vector $\alpha^*(i)$ are the corresponding components not equal: $\alpha_i = \frac{1}{i} > \alpha_{i+1} = 0$. Analogously, if there are several equally important criteria, then from extreme points corresponding to their indexes remains only the last one, because for others the corresponding components are not equal. If all the criteria are equally important, then all importance coefficients α_i are determined explicitly and the set $\bar{A}$ consists of the single point $\alpha^*(m) = (\frac{1}{m}, \ldots, \frac{1}{m})$. Denote $I^* = \{i_1, i_2, \ldots, i_n\}$ as the set of numbers $i \in \{1, \ldots, m\}$, for which there exist corresponding extreme points $\alpha^*(i)$ in accordance with described mapping. It means that there are n groups of equally important criteria, and numbers $i_1 < i_2 < \ldots < i_n = m$ are indexes of the last criteria in every group.

Assume that relation $yR^{\bar{A}^*}z$ holds for given arbitrary vector estimates y and z. It means that at all extreme points (6.4) of polyhedron $\bar{A}$, the values of the functions (6.3) do not exceed 0. In other words, for all numbers $i \in I^*$ and $k \in \{1, \ldots, q-1\}$, the following conditions hold:

$$\Delta_k(\alpha^*(i)) = \frac{1}{i}\sum_{j=1}^{i} c_{ik}(y,z) \leqslant 0,\ k = 1, \ldots, q-1. \tag{6.5}$$

According to (6.3), the numbers $c_{ik}(y,z)$ are determined as follows:

1. if $\sigma_i^k(y) = \alpha_i$, $\sigma_i^k(z) = \alpha_i$, then $c_{ik}(y,z) = 0$;
2. if $\sigma_i^k(y) = \alpha_i$, $\sigma_i^k(z) = 0$, then $c_{ik}(y,z) = 1$;
3. if $\sigma_i^k(y) = 0$, $\sigma_i^k(z) = \alpha_i$, then $c_{ik}(y,z) = -1$;
4. if $\sigma_i^k(y) = 0$, $\sigma_i^k(z) = 0$, then $c_{ik}(y,z) = 0$.

Hence, the condition (6.5) means that for the first i elements of vectors $\sigma^k(y)$ and $\sigma^k(z)$ total occurrences of the third case is not less than total occurrences of the second case.

We will fix an arbitrary number $k \in \{1, \ldots, q-1\}$ and show that the following vector inequality holds (see (3.2)):

$$\sigma_{\uparrow}^k(y) \leqq \sigma_{\uparrow}^k(z). \tag{6.6}$$

Assume that this is not the case and there exists $i \in \{1, \ldots, m\}$ such that $\sigma_{(i)}^k(y) > \sigma_{(i)}^k(z)$. It means that $\sigma_{(i)}^k(y) = \alpha_{i_p} > 0$, where $i_p \in I^*$, and $\sigma_{(i)}^k(z)$ are equal to either 0 or $\alpha_{i_r} > 0$, where $i_r \in I^*$ with a requirement of $r > p$. The numbers p and r are the indexes of groups of equally important criteria. Consequently, for the first i_p elements of vectors $\sigma^k(y)$ and $\sigma^k(z)$ the total occurrences of the third case is less than that of the second case, but this result contradicts our assumption.

Hence, vector inequality (6.6) holds. Since the number $k \in \{1, \ldots, q-1\}$ was chosen arbitrarily, all conditions of decision rule (3.2) are satisfied. Therefore relation $yR^{\Omega^*}z$ holds.

If relation $yI^{\mathrm{A}^*}z$ holds for vector estimates y and z, then $yR^{\mathrm{A}^*}z$ and $zR^{\mathrm{A}^*}y$. Therefore, as it was proved above, $yR^{\Omega^*}z$ and $zR^{\Omega^*}y$. Hence, $yI^{\Omega^*}z$ also holds.

If relation $yP^{\mathrm{A}^*}z$ holds for vector estimates y and z, then $yR^{\mathrm{A}^*}z$ is true and $zR^{\mathrm{A}^*}y$ is false. Therefore, as it was proved above, $yR^{\Omega^*}z$. And $zR^{\Omega^*}y$ is not true because of enclosure $R^{\Omega^*} \subseteq R^{\mathrm{A}^*}$ from (4.1). Hence, $yP^{\Omega^*}z$ also holds.

Enclosures from (4.1) together with derived enclosures

$$(R^{\mathrm{A}^*} \subseteq R^{\Omega^*},\ I^{\mathrm{A}^*} \subseteq I^{\Omega^*},\ P^{\mathrm{A}^*} \subseteq P^{\Omega^*})$$

complete the proof of Theorem 1.

Proof of Theorem 2.

Correctness of this theorem is shown by the following:

Counterexample. Assume that $m = 4$, $q = 3$, $\Omega^* = \{1 \succ 2,\ 1 \succ 3,\ 1 \succ 4,\ 2 \succ 3,\ 2 \succ 4,\ 3 \succ 4\}$. Systems of restrictions (4.2) and (4.3), which specify the set A^*, can be written in the compact form:

$$\alpha_1 > \alpha_2 > \alpha_3 > \alpha_4 > 0,\ \alpha_1 + \alpha_2 + \alpha_3 + \alpha_m = 1. \qquad (6.7)$$

Consider the vector estimates $y = (2, 2, 3, 1)$, $z = (1, 3, 2, 2)$. Taking into account (6.7), it can be shown that conditions from (3.9) hold:

$$\sup_{\alpha\in\mathrm{A}^*}\sum_{j=1}^{1}\sum_{i=1}^{m}(\sigma_i^j(y)-\sigma_i^j(z)) = \sup_{\alpha\in\mathrm{A}^*}(\alpha_4-\alpha_1) < 0,$$

$$\sup_{\alpha\in\mathrm{A}^*}\sum_{j=1}^{2}\sum_{i=1}^{m}(\sigma_i^j(y)-\sigma_i^j(z)) = \sup_{\alpha\in\mathrm{A}^*}[(\alpha_1+\alpha_2+2\alpha_4)-(2\alpha_1+\alpha_3+\alpha_4)] =$$

$$= \sup_{\alpha\in\mathrm{A}^*}[(\alpha_2-\alpha_1)+(\alpha_4-\alpha_3)] < 0.$$

Hence, $yR^{\mathrm{A}^*\&D}z$ is true.

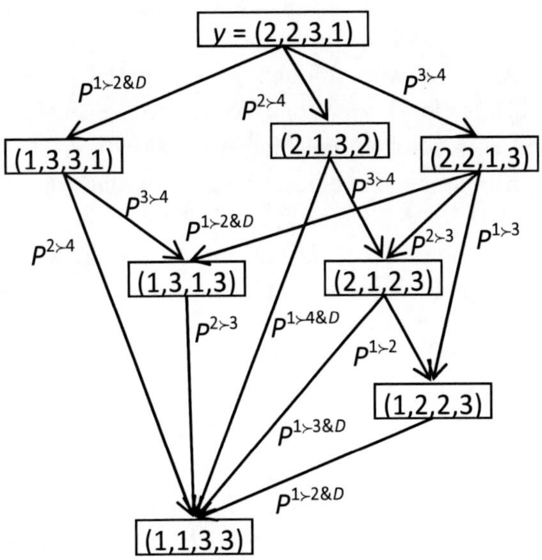

Figure 3. Vector estimates that are less preferable than y, according to the information $\Omega^*\&D$.

Let us check now $yR^{\Omega^*\&D}z$. It can be shown that it is impossible to construct the required sequence (3.1). Really, when $P^{i\succ j}$ or $P^{i\succ j\&D}$ are "applied" to a vector estimate, the sum of its components does not change, but in the case of P^0 this sum decreases. Since $2 + 2 + 3 + 1 = 1 + 3 + 2 + 2 = 8$, the sequence from y to z cannot contain any P^0. The graph in Figure 3 shows how the vector estimate $y = (2, 2, 3, 1)$ can be sequentially made worse, and the graph in Figure 4 shows how the vector estimate $z = (1, 3, 2, 2)$ can be sequentially made better.

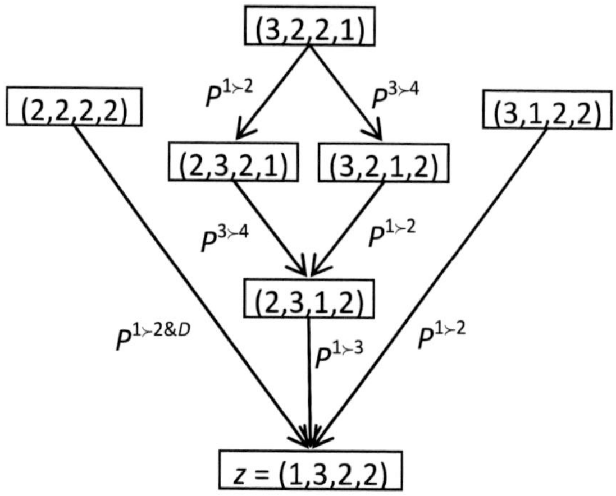

Figure 4. Vector estimates that are more preferable than z, according to the information $\Omega^*\&D$.

Since there is no equal vector estimates on these graphs, it is impossible to "stick together" their elements to form the sequence from y to z. Hence, $R^{\Omega^*\&D}$ is false.

References

[1] Belton V., Gear T. On the meaning of relative importance. *Journal of Multi-Criteria Decision Analysis*. 1997. V. 6. P. 335 – 338.

[2] Podinovski VV. An axiomatic solution of the problem of criteria importance in multicriterial problems. In: *Theory of Operations Research: State of the Art*. N.N. Moiseev (Ed.). Moscow: Nauka Publishing House. 1979. P. 117 –149 (in Russian).

[3] Podinovskii V.V. Multicriterial problems with importance-ordered criteria. *Automation and Remote Control*. 1976. V. 37. No 11, part 2. P. 1728 – 1736.

[4] Podinovskii V.V. Importance coefficients of criteria in decision making problems. Serial, or ordinal importance coefficients. *Automation and Remote Control*. 1978. V. 39. P. 1514 – 1524.

[5] Podinovski V.V. Problems with importance-ordered criteria // User-Oriented Methodology and Techniques of Decision Analysis and Support. J. Wessels and A.P. Wierzbicki (Eds.). *Lecture Notes in Economics and Math. Systems*. V. 397. Berlin: Springer, 1993. P. 150 – 155.

[6] Podinovski V.V. Multicriteria optimization problems involving importance-ordered criteria // *Modern Mathematical Methods of Optimization*. Elster K.-H. (Ed.). Berlin: Akademie Verlag. 1993. P. 254 – 267.

[7] Podinovski V.V. Criteria importance theory // *Mathematical Social Sciences*. 1994. V. 27. P. 237 – 252.

[8] Podinovski V.V. The quantitative importance of criteria for MCDA // *Journal of Multi-Criteria Decision Analysis*. 2002. V. 11. P. 1 – 15.

[9] Podinovski V.V. On the use of importance information in MCDA problems with criteria measured on the first ordered metric scale // *Journal of Multi-Criteria Decision Analysis*. 2009. V. 15. P. 163 – 174.

[10] Fishburn P.C. *Decision and value theory*. New York: Wiley, 1964.

[11] Podinovski V.V. Decision under multiple estimates for the importance coefficients of criteria and probabilities of values of uncertain factors in the aim function. *Automation and Remote Control*. 2004. V. 65. P. 1817 – 1833.

[12] Claessens M. N. A. J., Lootsma F. A., Vogt F. J. An elementary proof of Poelinck's theorem on the convex hull of ranked criterion weights // *European Journal of Operational Research*. 1991. V. 52. P. 255 – 258.

In: Several Problems of Applied Mathematics and Mechanics ISBN: 978-1-62081-603-5
Editors: Ivane Gorgidze and Tamar Lominadze

Chapter 5

ON THREE LAYER DIFFERENCE SCHEMES FOR SOLVING SYSTEMS OF MULTIDIMENSIONAL EQUATIONS OF A PARABOLIC TYPE WITH MIXED DERIVATIVES

***Francisco Criado*[1,*], *Tinatin Davitashvili*[2,†] *and Hamlet Meladze*[3,‡]**
[1]Malaga University, Malaga, Spain
[2]Tbilisi State University, Tbilisi, Georgia
[3]St. Andrew the First Called Georgian University, Tbilisi, Georgia

Abstract

The mixed problem with first order boundary conditions for systems of equations of parabolic types are considered:

$$B\frac{\partial u}{\partial t} = Lu + f,$$

where B - positively defined, symmetric, and bounded matrix, L - strong elliptic operator with variable coefficients, containing the mixed derivatives, $u = \left(u^{(1)}, u^{(2)}, \ldots, u^{(n)}\right)$ and $f = \left(f^{(1)}, f^{(2)}, \ldots, f^{(n)}\right)$ - n-dimensional vectors. The constructed three-layer factorized scheme's solution requires no inversion of matrix B. The received algorithms can be effectively used for multiprocessing computing systems. For difference schemes, the aprioristic estimation on layer in the norm of mesh space $\overset{0}{W}{}_2^{(1)}$ is received, on which the basis

* E-mail address: f_criado@uma.es
† E-mail address: t_davitashvili@hotmail.com
‡ E-mail address: h_meladze@hotmail.com

of the solution convergence of the difference scheme to the solution of an initial problem is proved.

1. Introduction

As it is known, the mathematical formulation of many scientific and engineering problems yield differential equations with partial derivatives of a parabolic type. Such problems are of concern: distribution of heat to the continuous environment, transport processes of substances, diffusion, movement of ground waters, dissipative biological structures, dynamics of power distribution in the hierarchy, etc. [1,2].

The solution of such problems is one of challenges in numerical mathematics and demands, as a rule, a lot of computing resources. One way to reduce the time for solving such problems is the use of parallel computing on multiprocessor systems.

A new quality of the structure of parallel computing systems requires a completely new mathematical framework for solving complex problems on these systems. This base is an adequate representation of the tasks in a form accessible to the parallel data processing. In computational mathematics, a large number of such methods are sufficiently developed. First of all, these are the methods of splitting complex problems into the simpler systems, the feeding realizations on the parallel processors.

The starting point in the given work is offered by the A.A. Samarskii method of regularization of difference schemes [3,4]. This method opens up great possibilities for constructing economical difference schemes of a given quality, which can be efficiently implemented on parallel computing systems.

In the present work, primary attention is paid to the following problems:

1. Getting the economic difference schemes with minimal requirements for a spatial operator. This requires only a strong ellipticity of the spatial operator.
2. Absolute stability at any $\tau < \infty$ (τ - a step on time) and $h < \infty$ (h_α - a step in a direction on x_α, $\alpha = 1,2,\ldots,p$).
3. Application of a one-dimensional double-sweep algorithm for the solution of received difference equations. These algorithms can be used for parallel computing systems.
4. Proof of convergence of difference schemes with less smoothness than the original system of differential equations, which is achieved by refraining from the local approximation.

Extensive literature is dedicated to the problems of difference splitting scheme construction. We note several monographs [3,5,6]. In these monographs, one can find an extensive list of works in this direction.

We also mention some works on parallel algorithms for solving parabolic equations [7-11]. Naturally, this list is incomplete.

2. Basic Concepts and Some Auxiliary Information

2.1. Let

$$D_p = (0,\ell_1)\times(0,\ell_2)\times\cdots\times(0,\ell_p) = \prod_{\alpha=1}^{p}(0,\ell_\alpha) =$$
$$= \{x = (x_1, x_2, \ldots, x_p),\ 0 < x_\alpha < \ell_\alpha,\ \alpha = 1,2,\ldots,p\}$$

-- open p–dimensional parallelepiped in p–dimensional Euclid space, E_p. Γ - surface of parallelepiped, and D_p and $\overline{D}_p = D_p \cup \Gamma$ -- is the D_p area's closure.

$G_T = D_p \times (0, T]$ -- cylinder in $(p+1)$-dimensional Euclid space with basis E_p and (x, t) – any point in G_T. $\overline{G}_T = \overline{D}_p \times [0, T]$ -- is the closure of the area G_T. $C^{\ell,k}(\overline{G}_T)$ -- the set of continuous $\overline{G}_T$ functions having $\overline{G}_T$ continuous derivatives up to an order ℓ (inclusive) on x and up to order k (inclusive) on t.

In cylinder $\overline{G}_T$, we introduce the difference grid, and the construction of the space-time grid in $\overline{G}_T$ is carried out by using one-dimensional grids on the intervals $[0,\ell_\alpha],\ \alpha = 1,2,\ldots,p$ and $[0, T]$:

$$\overline{\omega}_\alpha = \{x_\alpha^{(i_\alpha)} = i_\alpha h_\alpha,\ i_\alpha = 0,1,\ldots,N_\alpha,\ N_\alpha h_\alpha = \ell_\alpha\},\ \alpha = 1,2,\ldots,p.$$

$$\omega_\alpha = \{x_\alpha^{(i_\alpha)} = i_\alpha h_\alpha,\ i_\alpha = 1,\ldots,N_\alpha - 1,\ N_\alpha h_\alpha = \ell_\alpha\},\ \alpha = 1,2,\ldots,p.$$

The difference grid in parallelepiped $\overline{D}_p$ is constructed in the following way:

$$\overline{\omega}_h = \prod_{\alpha=1}^{p}\overline{\omega}_\alpha = \{x = (i_1 h_1,\ldots,i_p h_p) \in \overline{D}_p,\ i_\alpha = 0,1,\ldots,N_\alpha,\ N_\alpha h_\alpha = \ell_\alpha\}.$$

$\omega_h = \prod_{\alpha=1}^{p}\omega_\alpha$ -- the set of internal points of difference grid in D_p.

We denote by $\gamma_h = \overline{\omega}_h \setminus \omega_h \equiv \{x \in \Gamma\}$ the set of knots, belonging to the Γ boundary, which are called boundary knots. On the interval $[0, T]$, we introduce the difference grid:

$$\overline{\omega}_\tau = \{t_j = j\tau,\ j = 0,1,\ldots,k,\ k\tau = T\}.$$

Let $\Omega_{h\tau} = \omega_h \times \omega_\tau$ be the partial-time grid of internal knots of cylinder G_T and (x, t_j) is any knot of $\Omega_{h\tau}$. $\Gamma_{h\tau} = \gamma_h \times \overline{\omega}_\tau$ be the set of knots of the difference grid on the lateral surface of the cylinder $\overline{G}_T$ and (x', t_j) is the knot belonging to $\Gamma_{h\tau}$. Then:

$$\overline{\Omega}_{h\tau} = \Omega_{h\tau} \cup \Gamma_{h\tau} = \overline{\omega}_h \times \overline{\omega}_\tau$$

is the partial-time grid in $\overline{G}_T$.

2.2. We will deal with functions of the discrete argument, defined on knots of the difference grid called the grid functions.

We will consider a set of grid vector-functions $y = \left(y^{(1)}, y^{(2)}, \ldots, y^{(n)}\right)$, defined on $\overline{\Omega}_{h\tau}$. For them, the symbol $y^{(j)}$ indicates the value of the grid vector-functions in the knots (x, t_j).

We will use designations:

$$x^{(\pm 1_\alpha)} = \left(x_1, \ldots, x_{\alpha-1}, x_\alpha \pm h_\alpha, x_{\alpha+1}, \ldots, x_p\right),$$

$$y^{(\pm 1_\alpha)} = y\left(x^{(\pm 1_\alpha)}\right) = \left(y_1\left(x^{(\pm 1_\alpha)}\right), \ldots, y_n\left(x^{(\pm 1_\alpha)}\right)\right),$$

$$|h|^2 = \sum_{\beta=1}^{p} h_\alpha^2 .$$

We enter the difference relations:

$$y_{\bar{x}_\alpha} = \frac{y - y^{(-1_\alpha)}}{h_\alpha}, \quad y_{x_\alpha} = \frac{y^{(+1_\alpha)} - y}{h_\alpha},$$

$$\left(y_{\bar{x}_\alpha}\right)_{x_\beta} = y_{\bar{x}_\alpha x_\beta} = \frac{y^{(-1_\alpha)} - 2y + y^{(+1_\alpha)}}{h_\alpha^2} = \Lambda_\alpha^0 y,$$

$$-\Lambda_\alpha^0 y = A_\alpha^0 y, \quad A^0 = \sum_{\alpha=1}^{p} A_\alpha^0 .$$

For the grid vector-functions defined on $\overline{\Omega}_{h\tau}$, besides the specified designations, we will use still the following:

$$\hat{y} = y(x, t_{j+1}), \quad y = y(x, t_j), \quad \breve{y} = y(x, t_{j-1}),$$

$$y_{\bar{t}} = \frac{y - \check{y}}{\tau}, \quad y_t = \frac{\hat{y} - y}{\tau}, \quad y_{\bar{t}t} = \frac{\hat{y} - 2y + \check{y}}{\tau^2}, \quad y_{\overset{\circ}{t}} = \frac{\hat{y} - \check{y}}{2\tau}.$$

2.3. The set of the grid functions, defined on the grid $\overline{\omega}_p$, denoted by H, and its subset, consisting of grid functions, becoming zero (vanishing) on γ_p - through $\overset{\circ}{H}$.

On set $\overset{\circ}{H}$, we introduce the scalar product:

$$(y, v) = \sum_{i=1}^{n} \left(y^i, v^i\right),$$

where

$$\left(y^i, v^i\right) = \sum_{x \in \omega_p} y^i(x) v^i(x) H, \quad H = \prod_{\alpha=1}^{p} h_\alpha .$$

It is obvious, that this scalar product induces the norm $\|y\|_0^2 = (y, y)$.

This is denoted by $\omega_p^{+\alpha} = \omega_p \cup \gamma_\alpha^+$, where γ_α^+ -- set of knots on the γ_p border, at which $x_\alpha = l_\alpha$. $\gamma_{\alpha\beta}^+ (\alpha \neq \beta)$ -- set of knots on the γ_p border, at which $x_\alpha = l_\alpha$, $x_\beta = l_\beta$ and etc. Similarly, we will designate through γ_α^- -- set of knots of the γ_p border, at which $x_\alpha = 0$ and etc.

We introduce the notation $\omega_p^+ = \omega_p \cup \gamma_{1,2,\ldots,p}^+$. We also need the scalar product:

$$(y, v]_\alpha = \sum_{i=1}^{n} \left(y^i, v^i\right] \quad \left(y^i, v^i\right] = \sum_{x \in \omega_p^+} y^i(x) v^i(x) H.$$

Let's enter the mesh space of Sobolev $\overset{\circ}{W}{}_2^{(1)}(\omega_p)$, consisting of the functions from set $\overset{\circ}{H}$. The norm in this space is given by the following equality:

$$\|u\|_1^2 = \|u\|^2_{\overset{\circ}{W}{}_2^{(1)}} = \sum_{\alpha=1}^{p} \left\|u_{\bar{x}_\alpha}\right\|_0^2, \quad \left\|u_{\bar{x}_\alpha}\right\|_0^2 = \left(1, \left(u_{\bar{x}_\alpha}\right)^2\right]. \tag{1}$$

Note that $\overset{\circ}{W}{}_2^{(1)}$ -- is full Hilbert space under the norm (1).

2.4. Let A -- the linear self-conjugate operator in Hilbert space H and $A>0$. We define a new scalar product:

$$(y,v)_A=(Ay,v).$$

By this definition, H is transformed into a new Hilbert space, which we denote by H_A. Norm in the H_A will be denoted by $\|\cdot\|_A$:

$$\|u\|_A^2=(Au,u).$$

3. The Statement of Problem

In cylinder $\overline{G}_T$, we considered the following problem for systems of equations of a parabolic type with mixed derivatives:

Let us find a continuous solution of system.

Find a continuous solution in $\overline{G}_T$ of the system of equations:

$$\sum_{j=1}^{n} b_{ij}(x,t)\frac{\partial u^{(j)}}{\partial t}=\sum_{j=1}^{n}\sum_{\alpha,\beta=1}^{p}\frac{\partial}{\partial x_\alpha}\left(K_{\alpha,\beta}^{i,j}(x,t)\frac{du^{(j)}}{dx_\beta}\right)+f^i(x,t), \tag{2}$$

$$i=1,2,\ldots,n.$$

which satisfies the boundary conditions:

$$u^{(i)}(x',t)=g^{(i)}(x',t),\ \text{when}\ x'\in\Gamma\times[0,T] \tag{3}$$

and the initial conditions:

$$u^{(i)}(x,0)=u_0^{(i)}(x),\ \text{when}\ x\in\overline{D}_p,\ i=1,2,\ldots,n. \tag{4}$$

We assume that the following conditions are fulfilled:

I. Coefficients $K_{\alpha,\beta}^{i,j}(x,t)$ are Lipschitz-continuous with respect to t. Assume also, that matrixes from coefficients of the spatial operator $K_{\alpha\beta}=\left(K_{\alpha,\beta}^{i,j}\right)_{i,j=1}^{n}$ $(\alpha,\beta=1,2,\ldots,p)$ are symmetric and that the problem (2)-(4) has the unique solution $u\equiv u(x,t)$, continuous in $\overline{G}_T$ and differentiable a necessary number of times.

II. The spatial operator of the system of equations (2) is the strong-elliptic, so that for any $t\in[0,T]$, the following inequality is true:

$$\nu_1 \sum_{\alpha=1}^{p} \sum_{i=1}^{n} \left(\xi_\alpha^i\right)^2 \le \sum_{\alpha,\beta}^{1\div p} \sum_{i,j}^{1\div n} K_{\alpha\beta}^{ij}\, \xi_\beta^j\, \xi_\alpha^i \le \nu_2 \sum_{\alpha=1}^{p} \sum_{i=1}^{n} \left(\xi_\alpha^i\right)^2, \tag{5}$$

where $\left\{\xi_\alpha^i\right\}$ -- any real numbers, ν_1 and ν_2 -- positive constants.

III. Matrix $B = \left(b_{ij}\right)$ positive-defined, symmetric, and bounded;

$$\|B\| \le \nu_3,\ B \ge \delta E,$$

where ν_3, δ -- positive constants and E - unit matrix.

4. Three-layer Difference Scheme for Problem (2)-(4)

4.1. For problem (2)-(4) we consider a three-layer difference scheme, which order of approximation in a class $C^{4,2}\left(\overline{G}_T\right)$ of solutions (2) is a quantity $O\left(\tau + |h|^2\right)$:

$$By_{\bar{t}} + \tau^2 R y_{\bar{t}t} + Ay = f, \quad (x,t) \in \Omega_{h\tau}, \tag{6}$$

where $y = \left(y^{(1)}, y^{(2)}, \ldots, y^{(n)}\right)$ – n-dimensional vector,

$$Ay = -\frac{1}{2} \sum_{\alpha,\beta}^{1\div p} \left[\left(K_{\alpha\beta}(x,t) y_{\bar{x}_\beta}\right)_{x_\alpha} + \left(K_{\alpha\beta}(x,t) y_{x_\beta}\right)_{\bar{x}_\alpha} \right], \tag{7}$$

R – operator- regularizer, the choice of which provides the absolute stability of the difference scheme (6).

The vector function $y(x,t)$ satisfies the following boundary and initial conditions:

$$\left.\begin{array}{l} y(x',t) = g(x',t), \quad \text{если} \quad x' \in \Gamma_{h\tau} \\ y(x,0) = u_0(x), \quad y(x,\tau) = u_1(x) \end{array}\right\}, \tag{8}$$

where the second initial condition on the layer $t=\tau$ can be approximated, for example, believing that $u_1(x) = u_0(x)$. The approximation error in this case is the quantity of order *O(τ)*.

We write the difference scheme (6) in canonical form [3]:

$$By_{\overset{\circ}{t}} + \tau^2 \left(R - \frac{1}{2\tau} B \right) y_{\bar{t}t} + Ay = f. \tag{9}$$

In work [4], the fairness of following relations is proved:

$$\nu_1\left(A^0 y, y\right) \le (Ay, y) \le \nu_2\left(A^0 y, y\right),$$

or

$$\nu_1 A^0 \le A \le \nu_2 A^0$$

for any vector $y \in H_0$, where the operator A^0 is defined above, in 2.2.

Consequently, the operators A and A^0 in H_0 are energetically equivalent with constants ν_1 and ν_2.

In paper [3] it is proved that the following three layer difference scheme:

$$By_{\bar{t}} + \tau^2 R y_{\bar{t}t} + Ay = \varphi(t), \quad (x,t) \in \Omega_{h\tau}, \; y(0) = y_0, \quad y(\tau) = y_1,$$
$$0 < t = n\tau < t_0, \quad n = 1,2,\ldots,n_0 - 1, \quad t_0 = n_0\tau,$$

is absolutely stable with respect to the initial data, and on the right side, provided that the given vectors y_0 and y_1 are the arbitrary vectors of finite-dimensional space H (H - Hilbert space), $\varphi(t)$ - given arbitrary abstract function, $t \in \overline{\omega}_\tau$, *A, B, and R* are linear operators on *H*. In addition:

$A = A(t) = A^*(t) > 0$, $R = R(t) = R^*(t) > 0$ - Lipschitz-continuous in t,

$B = B(t) > \delta E$ (where δ is a positive constant, and E - identity operator),

$R(t) = \dfrac{1+\varepsilon}{4} A(t)$ for all $0 < t = n\tau < t_0$ (where $\varepsilon = const$ does not depend on τ and h).

Based on this result, we can conclude that the difference scheme (9) with regularizer:

$$R = \sigma A^0 + \frac{\nu_3}{2\tau} E, \tag{10}$$

at $\sigma \ge \dfrac{1+\varepsilon}{4}\nu_2$ $(0 < \varepsilon \le 1)$ is absolutely stable in space H_0 (stable with respect to initial data and on the right part). So, we will assume that operator R is defined by means of equality (10).

4.2. Proceed to the construction of the economic factorized difference scheme, considering scheme (6) as the initial.

We write the difference scheme (6) as follows:

$$By_{\bar{t}} + \frac{\nu_3 \tau}{2}\left(E + \tau R^0\right) y_{\bar{t}t} + Ay = f, \tag{11}$$

where

$$R^0 = \overline{\sigma} A^0 = \sum_{\alpha=1}^{p} R_\alpha^0, \qquad R_\alpha^0 = \overline{\sigma} A_\alpha^0, \qquad \overline{\sigma} = \frac{2\sigma}{\nu_3}.$$

Equation (11) can be written as follows:

$$\frac{\nu_3 \tau}{2}\left(E + \tau R^0\right) y_{\bar{t}t} = f - Ay - By_{\bar{t}}. \tag{12}$$

Replacing the operator $E + \tau R^0 = E + \tau \sum_{\alpha=1}^{p} R_\alpha^0$ with the factorized operator $\prod_{\alpha=1}^{p}\left(E + \tau R_\alpha^0\right)$, we receive the three-layer factorized scheme:

$$\frac{\nu_3 \tau}{2} \prod_{\alpha=1}^{p}\left(E + \tau R_\alpha^0\right) y_{\bar{t}t} = f - Ay - By_{\bar{t}}. \tag{13}$$

We use the equality:

$$\prod_{\alpha=1}^{p}\left(E + \tau R_\alpha^0\right) = E + \tau \sum_{\alpha=1}^{p} R_\alpha^0 + \tau^2 Q_p,$$

where:

$$Q_p = \sum_{\alpha<\beta} R_\alpha^0 R_\beta^0 + \tau \sum_{\alpha<\beta<\gamma} R_\alpha^0 R_\beta^0 R_\gamma^0 + \cdots + \tau^{p-2} \prod_{\alpha=1}^{p} R_\alpha^0,$$

then the difference scheme (12) can be written as follows:

$$By_{\bar{t}} + \frac{\nu_3 \tau}{2}\left(E + \tau R^0 + \tau^2 Q_p\right) y_{\bar{t}t} + Ay = f, \tag{14}$$

or, in canonical form:

$$By_{\overset{\circ}{t}} + \tau^2 \widetilde{R}\, y_{\bar{t}t} + Ay = f, \tag{15}$$

where

$$\widetilde{R} = \frac{\nu_3}{2} R^0 + \frac{\nu_3}{2\tau} E - \frac{1}{2\tau} B + \frac{\nu_3 \tau}{2} Q_p.$$

It is evident that if $\sigma \geq \frac{1+\varepsilon}{4}\nu_2$ $(0<\varepsilon \leq 1)$, then $\widetilde{R} \geq \frac{1+\varepsilon}{4}A$ and factorized scheme (15) will be absolutely stable in H_0.

4.3. Note that for difference schemes (13), we can construct an efficient numerical algorithm using the decomposition of the operator $\prod_{\alpha=1}^{p}\left(E+\tau R_\alpha^0\right)$.

The difference scheme (13) we rewrite as follows:

$$\prod_{\alpha=1}^{p} B_\alpha y_{\bar{t}t} = F,$$

where $B_\alpha = E + \tau\bar{\sigma}A_\alpha^0$, $F = \frac{2}{\nu_3 \tau}\left(f - Ay - By_{\bar{t}}\right)$.

Such a record of difference schemes gives the opportunity to construct the following computational algorithm:

$$\begin{gathered} B_1 v_{(1)} = F, \quad B_\alpha v_{(\alpha)} = v_{(\alpha-1)} \quad (\alpha = 2,3,\dots,p), \qquad v_{(p)} = y_{tt} \\ y^{j+1} = \tau^2 v_{(p)} + 2y^j - y^{j-1}, \quad j = 1,2,3\dots \end{gathered}, \tag{16}$$

For definitions of the functions $v_{(\alpha)}$ at $x \in \Gamma_{h\tau}$, it is necessary to use the following formulas:

$$v_{(\alpha)} = B_{\alpha+1} \cdots B_p g_{tt}(x,t) \text{ при } x \in \gamma_\alpha^+ \cup \gamma_\alpha^- \cup [0,T]. \tag{17}$$

The obtained problems (16)-(17) can be solved by applying one of the one-dimensional factorization algorithms (Thomas algorithm) [12]. This algorithm is particularly suitable for solving the considered problem in the case, when the function $g(x,t)$ does not depend on t. In this case, we have homogeneous boundary conditions for functions $v_{(\alpha)}$ $(\alpha = 1,2,\dots,p)$.

Thus, the original problem (13) splits into several subtasks, each of which can be solved using parallel computing algorithms [13-15].

5. Convergence of the Difference Scheme

5.1. During construction of the initial difference scheme (6), we demanded that the solution of the differential equation systems (2) belongs to the class $C^{4,2}(\overline{G}_T)$. An approximation error of factorized scheme (15) is presented in the form $\psi = \psi_0 + \psi_1$, where ψ_0 -- the error of approximation of the original scheme (6), the order of which in the considered class is equal to $O\left(\tau+|h|^2\right)$, and $\psi_1=\tau^3\frac{\nu_3}{2}Q_p u_{\bar{t}t}$.

It follows that if $p>2$, the requirement $\psi_1=O\left(\tau+|h|^2\right)$ imposes additional restrictions on the smoothness of solutions of the system of equations (2). However, the aprioristic estimation and the theorem of convergence in the norm of space $\overset{\circ}{W}{}_2^{(1)}$ can be obtained under weaker restrictions than the condition of local approximation of the order $O\left(\tau+|h|^2\right)$.

5.2. To prove the convergence of factorized scheme (15), we consider the vector function $z=y-u,\ (x,t)\in\overline{\Omega}_{h\tau}$, where u -- the solution of initial problem (2)-(4), and y - the solution of the difference scheme (15), (8).

For grid function z, we receive the following problem:

$$\left.\begin{aligned} &Bz_{\overset{\circ}{t}}+\tau^2\widetilde{R}z_{\bar{t}t}+Az=\psi,\\ &z(0)=0,\quad z(\tau)=\tau\,\mu, \end{aligned}\right\} \tag{18}$$

where $\psi=\psi_0+\psi_1$ -- the error of approximation of difference scheme (15), and $|\mu(x)|\le\nu_4=const$.

We consider $z=z(t)$ as an abstract function of time argument $t\in\overline{\omega}_\tau$ with values in H_0, where H_0 – set of the vector functions defined on $\overline{\omega}_p$ and equal to zero on the border γ_h of the grid $\overline{\omega}_p$.

To obtain a priori estimation, it is convenient to split the problem (18) into two tasks:

$$\left.\begin{aligned} &Bz_{\overset{\circ}{t}}+\tau^2\widetilde{R}z_{\bar{t}t}+Az=0,\\ &z(0)=0,\quad z(\tau)=\tau\,\mu, \end{aligned}\right\} \tag{19}$$

and

$$\left.\begin{aligned} &Bz_{\overset{\circ}{t}}+\tau^2\widetilde{R}z_{\bar{t}t}+Az=\psi,\\ &z(0)=0,\quad z(\tau)=0, \end{aligned}\right\} \tag{20}$$

The solution to problem (18) can be represented as a sum $z = z_1 + z_2$, where z_1 -- solution of problem (19) and z_2 -- solution of problem (20). It is evident that $z_1 \in H_0$ and $z_2 \in H_0$.

5.3. To obtain a priori estimation, we will use the energetic identity corresponding to factorized difference scheme (4.1). We will take the scalar product of both members of the equation (18) on $2\tau z_{\overset{\circ}{t}} = \tau(z_t + z_{\bar{t}}) = \hat{z} - \breve{z}$. Taking into account the form of the operator $\widetilde{R}$ and the relation $\tau z_{\bar{t}t} = z_t - z_{\bar{t}}$, we will receive:

$$2\tau\left(Bz_{\overset{\circ}{t}},\, z_{\overset{\circ}{t}}\right) + \tau^2\left(\left(\frac{\nu_3}{2\tau}E - \frac{1}{2\tau}B\right)(z_t - z_{\bar{t}}), (z_t + z_{\bar{t}})\right) +$$
$$+\tau^2\left(\left(\frac{\nu_3}{2}R^0 + \frac{\nu_3}{2}\tau Q_p - \frac{1}{2}A\right)(z_t - z_{\bar{t}}), (z_t + z_{\bar{t}})\right) +$$
$$+\frac{1}{2}\left(A(\hat{z} + \breve{z}), \hat{z} - \breve{z}\right) = 2\tau\left(\psi, z_{\overset{\circ}{t}}\right).$$

After some simple transformations we obtain:

$$2\tau\left(Bz_{\overset{\circ}{t}},\, z_{\overset{\circ}{t}}\right) + \tau^2\left(\left(\frac{\nu_3}{2\tau}E - \frac{1}{2\tau}B\right)(z_t - z_{\bar{t}}),\ z_t + z_{\bar{t}}\right) + \frac{1}{4}\left(A(\hat{z} + z), \hat{z} + z\right) +$$
$$+\tau^2\left(\left(\frac{\nu_3}{2}R^0 - \frac{1}{4}A\right)z_t,\ z_t\right) + \tau^3\frac{\nu_3}{2}\left(Q_p z_t,\ z_t\right) =$$
$$= \frac{1}{4}\left(A(z + \breve{z}), z + \breve{z}\right) + \tau^2\left(\left(\frac{\nu_3}{2}R^0 - \frac{1}{4}A\right)z_{\bar{t}},\ z_{\bar{t}}\right) +$$
$$+\tau^3\frac{\nu_3}{2}\left(Q_p z_{\bar{t}},\ z_{\bar{t}}\right) + 2\tau\left(\psi, z_{\overset{\circ}{t}}\right).$$

The last equality allows us to obtain the energetic identity, which in the case of problem (18) has the form:

$$2\tau\left(Bz_{\overset{\circ}{t}},\, z_{\overset{\circ}{t}}\right) + \|\hat{z}\|^2_{(2)} = \|z\|^2_{(2)} + \tau F + 2\tau\left(\psi, z_{\overset{\circ}{t}}\right), \tag{21}$$

where:

$$\|\hat{z}\|^2_{(2)} = \|z(t+\tau)\|^2_{(2)} = \frac{1}{4}\|\hat{z} + z\|^2_A + \tau^2\left(\left(\frac{\nu_3}{2}R^0 - \frac{1}{4}A\right)z_t,\ z_t\right) +$$
$$+\tau^3\frac{\nu_3}{2}\left(Q_p z_t,\ z_t\right) + \tau^2\left(\left(\frac{\nu_3}{2\tau}E - \frac{1}{2\tau}B\right)z_t, z_t\right),$$

$$\|z\|_{(2)}^{2}=\|z(t)\|_{(2)}^{2}=\frac{1}{4}\|z+\breve{z}\|_{\breve{A}}^{2}+\tau^{2}\left(\left(\frac{\nu_3}{2}R^{0}-\frac{1}{4}\breve{A}\right)z_{\bar{t}},\ z_{\bar{t}}\right)+$$
$$+\frac{\tau^{3}\nu_3}{2}\left(Q_p z_{\bar{t}},\ z_{\bar{t}}\right)+\tau^{2}\left(\left(\frac{\nu_3}{2\tau}E-\frac{1}{2\tau}B\right)z_{\bar{t}},z_{\bar{t}}\right),$$

$$F=\frac{1}{4}\left(A_{\bar{t}}\left(z+\hat{z}\right),z+\breve{z}\right)+\tau^{2}\left(\left(\frac{\nu_3}{2}R^{0}-\frac{1}{4}A\right)_{\bar{t}}z_{\bar{t}},\ z_{t}\right)$$

and $\left(Q_p\right)_{\bar{t}}=0$, as the operator Q_p does not depend on t.

We use the condition of absolute stability of the difference scheme in space H_0, then:

$$R=\sigma A^{\circ}+\frac{\nu_3}{2\tau}E\geq\frac{1+\varepsilon}{4}A.$$

This condition is valid if:

$$R=\sigma A^{\circ}+\frac{\nu_3}{2\tau}E\geq\frac{1+\varepsilon}{4}A.\ \text{or}\ \frac{\nu_3}{2}R^{0}\geq\frac{1+\varepsilon}{4}A\ \left(R^{0}=\bar{\sigma}A^{0},\ \bar{\sigma}=\frac{2\tau}{\nu_3}\right).\qquad(22)$$

The operator $A=A(t)$ is self-conjugate, positive, and Lipschitz- continuous with respect to t:

$$\begin{aligned}&A^{*}(t)=A(t)>0,\\&\left|\left(A_t y,y\right)\right|\leq\nu_5\left|\left(\breve{A}y,y\right)\right|\quad\text{или}\quad\left|\left(Ay,y\right)\right|\leq\left(1+\nu_5\tau\right)\left(\breve{A}y,y\right),\quad\breve{A}=A(t-\tau),\end{aligned}\qquad(23)$$

where ν_5 -- the Lipschitz constant for the operator A.

Further, if we use a technique from [16], and relations (22) and (23), we obtain the following estimation:

$$|F|\leq\nu_5\|z\|_{(2)}^{2}.$$

This allows us to write the energetic inequality:

$$2\tau\left(Bz_{\overset{\circ}{t}},z_{\overset{\circ}{t}}\right)+\|\hat{z}\|_{(2)}^{2}\leq\left(1+\nu_5\tau\right)\|z\|_{(2)}^{2}+2\tau\left(\psi,z_{\overset{\circ}{t}}\right),\qquad(24)$$

based on which we can obtain the desired priori estimate.

5.4. For further presentation, it is useful to introduce the following **notations**:

$$A_\alpha^0 = T_\alpha T_\alpha^* \text{ or } A_\alpha^0 u = T_\alpha T_\alpha^* u = u_{\bar{x}_\alpha x_\alpha}, \quad T_\alpha u = u_{\bar{x}_\alpha}, \quad T_\alpha^* = u_{x_\alpha},$$

$$\|y\|_{q_{2,p}}^2 = \sum_{\alpha<\beta}^{1\div p} \|T_\alpha T_\beta y\|_0^2, \quad \|y\|_{q_{3,p}}^2 = \sum_{\alpha<\beta<\gamma}^{1\div p} \|T_\alpha T_\beta T_\gamma y\|_0^2, \ldots, \|y\|_{q_{p,p}}^2 = \sum_{\alpha<\beta}^{1\div p} \|T_1 T_2 \cdots T_p y\|_0^2. \tag{25}$$

Theorem 1. If conditions I-III are satisfied, regularizer R is defined by equality (10) and $\sigma \geq \frac{1+\varepsilon}{4}\nu_2$ $(\varepsilon = const > 0)$. Then for the solution z_1 of the problem (19) on any sequence of grids $\overline{\Omega}_{h\tau}$, the aprioristic estimation is fair:

$$\left\|z_1^{j+1}\right\|_{(2)} \leq M_1 \tau \left\{ \|\mu(x)\|_0^2 + \tau \|\mu(x)\|_{A^0}^2 + \sum_{s=2}^{p} \tau^s \|\mu(x)\|_{q_{s,p}}^2 \right\}^{1/2}, \tag{26}$$

where $M_1 > 0$ -- the constant number which is not dependent on the grid, and the norms $\|\mu(x)\|_{q_{s,p}}^2$ are defined by equalities (25).

The proof. Given the reasoning and similarity to [16] (the theorem 1), we receive the inequality expressing stability of the factorized scheme (19) with respect to initial data:

$$\left\|z_1^{j+1}\right\|_{(2)} \leq M_1' \|z(\tau)\|_{(2)}, \quad j = 1,2,\ldots,$$

where $M_1' > 0$ -- a constant which is not dependent on the mesh.

Let us transform the right part of the given inequality:

$$\|z(\tau)\|_{(2)}^2 = \frac{1}{4}\|\tau\,\mu(x)\|_{\breve{A}}^2 + \left(\left(\frac{\nu_3}{2}R^0 - \frac{1}{4}\breve{A}\right)\tau\,\mu(x),\ \tau\,\mu(x)\right) +$$

$$+ \frac{\tau^3 \nu_3}{2}(Q_p\,\mu(x),\ \mu(x)) + \tau^2\left(\left(\frac{\nu_3}{2\tau}E - \frac{1}{2\tau}B\right)\ \mu(x), \mu(x)\right) \leq$$

$$\leq \frac{\nu_3}{2}\tau^2\left(R^0 \mu(x), \mu(x)\right) + \frac{\tau^3 \nu_3}{2}(Q_p\,\mu(x),\ \mu(x)) + \tau\frac{\nu_3}{2}(\mu(x), \mu(x)).$$

Further, to convert an expression $(Q_p\,\mu(x),\ \mu(x))$, we use Green's first difference formula [3], considering that $\mu(x) \in H_0$. From this follows the fairness of inequality (26).

The theorem is therefore proved.

Further required is the following:

Lemma. If $a > 0$, $0 < b < \min\left(\frac{1}{T}, 1\right)$, $a = const$, $b = const$ and $0 \le \tau \le T$, then:

$$\frac{1 + a\tau}{1 - b\tau} \le 1 + k\tau , \tag{27}$$

where:

$$k \ge \frac{a + b}{1 - b\tau} . \tag{28}$$

Proof. From inequality (27), the following can be obtained:

$$k \ge \frac{a + b}{1 - b\tau} .$$

The smallest value of the denominator of this fraction is at $\tau = T$. Thus, $\max k$ is reached at $\tau = T$, and if k is defined by (28), the inequality (27) will be valid for any $0 < \tau \le T$.

Theorem 2. Let conditions I-III be satisfied. In addition, the regularizer R is defined as follows:

$$R = \sigma A_0 + \frac{\nu_3}{2\tau} E ,$$

where $\sigma \ge \frac{1 + \varepsilon}{4} \nu_2$ $0 < \varepsilon < \min\left(1, \frac{1}{T}\right)$, ε does not depend on the grid. Then on any sequence of grids for the solution z_2 of the problem (20), the aprioristic estimation is fair:

$$\left\| z_2^{j+1} \right\|_2 \le M_2 \max_{0 < t' \le t_j} \left\| \psi_0 (t') \right\|_0 + M_3 \tau \max_{0 < t' \le t_j} \left\{ \sum_{s=2}^{p} \tau^{s-2} \left\| u_{\bar{t}} (t') \right\|_{q_{s,p}}^2 \right\}^{1/2} , \tag{29}$$

where $M_2 > 0$, $M_3 > 0$ -- constants, which depends only on $\varepsilon, \nu_1, \nu_2, \nu_5, T$, and area $\overline{D}_p$.

Proof. We rewrite the energetic inequality (4.9) in a form:

$$2\tau \left(B z_{\overset{\circ}{t}}, z_{\overset{\circ}{t}} \right) + \left\| \hat{z} \right\|_{(2)}^2 \le (1 + \nu_5 \tau) \left\| z \right\|_{(2)}^2 + 2\tau \left(\psi_0, z_{\overset{\circ}{t}} \right) + 2\tau \left(\psi_1, z_{\overset{\circ}{t}} \right) . \tag{30}$$

To obtain the aprioristic estimation, we transform the expression $2\tau\left(\psi_1, \overset{\circ}{z}_t\right)$. Using Green's first difference formula and ε–inequality [3], we receive:

$$2\tau\left(\psi_1, \overset{\circ}{z}_t\right) = \frac{\nu_3}{2}\tau^4\left\{\left(Q_p u_{\bar{t}t}, z_t\right) + \left(Q_p u_{\bar{t}t}, z_{\bar{t}}\right)\right\} \le$$
$$\le M_2'(\bar{\sigma})\frac{\nu_3}{2}\tau^4\frac{1}{2\varepsilon}\left(\sum_{s=2}^{p}\tau^{s-2}\left\|u_{\bar{t}t}\right\|^2_{q_{s,p}}\right) + \frac{\nu_3}{2}\tau^4\varepsilon\left\{\left(Q_p z_t, z_t\right) + \left(Q_p z_{\bar{t}}, z_{\bar{t}}\right)\right\} \le$$
$$\le M_2'(\bar{\sigma})\frac{\nu_3}{4\varepsilon}\tau^4\left(\sum_{s=2}^{p}\tau^{s-2}\left\|u_{\bar{t}t}\right\|^2_{q_{s,p}}\right) + \varepsilon\tau\left\|\hat{z}\right\|^2_{(2)} + \varepsilon\tau\left\|z\right\|^2_{(2)} \le$$
$$\le M''\tau^3\left(\sum_{s=2}^{p}\tau^{s-2}\left\|u_{\bar{t}}\right\|^2_{q_{s,p}}\right) + \varepsilon\tau\left\|\hat{z}\right\|^2_{(2)} + \varepsilon\tau\left\|z\right\|^2_{(2)}. \tag{31}$$

where M_2'' is a constant and does not depend on the grid.

On the other hand, the inequality is fair:

$$2\tau\left(\psi_0, \overset{\circ}{z}_t\right) \le 2\tau\delta\left\|\overset{\circ}{z}_t\right\|^2_0 + \frac{\tau}{2\delta}\left\|\psi_0\right\|^2_0. \tag{32}$$

Substituting estimations (31) and (32) in inequality (30), we receive:

$$(1 - \varepsilon\tau)\left\|\hat{z}_2\right\|^2_{(2)} \le \left(1 + (\nu_5 + \varepsilon)\tau\right)\left\|z\right\|^2_{(2)} + \frac{\tau}{2\delta}\left\|\psi_0\right\|^2_0 + M_2''\tau^3\sum_{s=2}^{p}\tau^{s-2}\left\|u_{\bar{t}}(t)\right\|^2_{q_{s,p}}. \tag{33}$$

Using Lemma, inequality (33) can be rewritten as follows:

$$\left\|\hat{z}_2\right\|^2_{(2)} \le (1 + \nu_6\tau)\left\|z\right\|^2_{(2)} + \tau\nu_7\left\|\psi_0\right\|^2_0 + \nu_8\tau^3\sum_{s=2}^{p}\tau^{s-2}\left\|u_{\bar{t}}(t)\right\|^2_{q_{s,p}},$$

where ν_6, ν_7, ν_8 -- the positive constants which are not dependent on the grid.

Sum the last inequality by $t' = 2\tau, 3\tau, \cdots, T$ and apply the difference analogue of Grönwall's lemma [16], on which basis a priori estimation (29) is obtained.

The theorem is therefore proved.

Uniting the results of theorems 1 and 2, we will prove that the following theorem is fair.

Theorem 3. Let conditions of theorem 2 be satisfied. Then, on any sequence of grids $\overline{\Omega}_{h\tau}$ for the solution of problem (18), the aprioristic estimation is fair:

$$\left\|z^{j+1}\right\|_{(2)} \le M_1\tau\left\{\left\|\mu(x)\right\|_0^2 + \tau\left\|\mu(x)\right\|_{A^0}^2 + \sum_{s=2}^{p}\tau^s\left\|\mu(x)\right\|_{q_{s,p}}^2\right\}^{1/2} +$$

$$+ M_2 \max_{0<t'\le t_j}\left\|\psi_0(t')\right\|_0 + M_3\tau \max_{0<t'\le t_j}\left\{\sum_{s=2}^{p}\tau^{s-2}\left\|u_{\bar t}(t')\right\|_{q_{s,p}}^2\right\}^{1/2},$$

where M_1, M_2, M_3 -- positive constants which do not depend on the grid.

Considering an estimation $\left\|z^{j+1}\right\|_{(2)} \ge \nu_9\left\|z^{j+1}\right\|_1$, where $\nu_9 > 0$ -- a constant, which is not dependent on the grid, on the basis of theorem 3 it is easily possible to receive an aprioristic estimation for the solution of problem (18) in space $\overset{\circ}{W}{}_2^{(1)}$.

Theorem 4. Let conditions of theorem 2 be satisfied. Then, on any sequence of meshes $\overline{\Omega}_{h\tau}$ for the solution of problem (18), the aprioristic estimation is fair:

$$\left\|z^{j+1}\right\|_{(1)} \le M_1'\tau\left\{\left\|\mu(x)\right\|_0^2 + \tau\left\|\mu(x)\right\|_{A^0}^2 + \sum_{s=2}^{p}\tau^s\left\|\mu(x)\right\|_{q_{s,p}}^2\right\}^{1/2} +$$

$$+ M_2' \max_{0<t'\le t_j}\left\|\psi_0(t')\right\|_0 + M_3'\tau \max_{0<t'\le t_j}\left\{\sum_{s=2}^{p}\tau^{s-2}\left\|u_{\bar t}(t')\right\|_{q_{s,p}}^2\right\}^{1/2},$$

where M_1', M_2', M_3' -- positive constants, which are not depend on mesh.

Theorem 5. Let conditions of theorem 2 be satisfied, and let solutions of the system of equations (2) $u \in C^{4,2}$, when $p \le 4$, and $u \in C^{p,2}$ when $p \ge 4$. Then the solutions of difference problem (15) and (8) converge in norm $\overset{\circ}{W}{}_2^{(1)}$ with speed $O\left(\tau + |h|^2\right)$ to the solution of the problem (2)-(4).

References

[1] A.A. Samarskii, A.N. Tikhonov. *Equations of Mathematical Physics*. Publisher: Dover Pubns, 1990, 766 p.

[2] A.P. Mikhailov, A.A. Samarskii. *Matematicheskoe modelirovanie: Idei, Metody, Primery* (Russian). Publisher: Fiziko-matematicheskaia literatura, 2005, 316 p. (A.P. Mikhailov, A.A. Samarskii. *Principles of Mathematical Modeling: Ideas, Methods, Examples.* Publisher: CRC Pr-ILIC, Numerical insights, 3, 2002).

[3] A.A. Samarskii. *The Theory of Difference Schemes. Publisher*: Narcel Dekker inc., 2001, 761 p.

[4] A.A. Samarskii. About regularization of difference schemes // *J. comp. math. and math. phys.* (A.A.Samarsky. Regularization of difference schemes. Zh. Math. and Math. nat.), 1967, 7, №1. pp. 62-93.

[5] G.I. Marchuk. *Numerical Methods and Applications*. Publ.: CRC Press, 1994, 288 p.

[6] A.A. Samarskii, P.N. Vabishevich, P.P. Matus. *Difference Schemes with Operator Factors*. Publ.: Kluwer Academic Pub., 2002, 396 c.

[7] Wenrui Hao a; Shaohong Zhu a. Parallel iterative methods for parabolic equations // *International Journal of Computer Mathematics*, Volume 86, Issue 3 March 2009, pages 431 – 440.

[8] R. K. Mohanty, D. J. Evans. Alternating group explicit parallel algorithms for the solution of one-space dimensional non-linear singular parabolic equations using an $O(k^2 + h^4)$ difference method // *International Journal of Computer Mathematics*, Volume 82, Issue 2 February 2005, pages 203 – 218.

[9] A. Q. M. Khaliq, E. H. Twizell, D. A. Voss. On parallel algorithms for semidiscretized parabolic partial differential equations based on subdiagonal Padé approximations // *Numerical Methods for Partial Differential Equations*, Volume 9 Issue 2 (March 1993). Pages 107 – 116 (Published Online: 20 Jun 2005).

[10] M.S.A. Taj; E.H. Twizell. A family of third-order parallel splitting methods for parabolic partial differential equations // *International Journal of Computer Mathematics*, 1029-0265, Volume 67, Issue 3, 1998, Pages 411 – 433.

[11] G. Horton, S. Vandewalle, and P. Worley. An Algorithm with Polylog Parallel Complexity for Solving Parabolic Partial Differential Equations // *SIAM J. Sci. Comput.* Volume 16, Issue 3, pp. 531-541 (May 1995).

[12] A. Quarteroni, R. Sacco, F. Saleri, *Numerical Mathematics*, Springer, 2007. p.669.

[13] J. Verkaik, H.X. Lin. A class of novel parallel algorithms for the solution of tridiagonal systems // *Parallel Computing*, Volume 31, Issue 6, June 2005, Pages 563-587.

[14] Xian-He Sun, Hong Zhang Sun, Lionel M. Ni. Parallel algorithms for solution of tridiagonal systems on multicomputers//*Proceedings of the 3rd international conference on Supercomputing*, Crete, Greece, 1989. Publisher: ACM, ISBN:0-89791-309-4, pp: 303 – 312.

[15] Paul F. Fischer, Franco P. Preparata, John E. Savage. Generalized scans and tri-diagonal systems. Book Series "*Lecture Notes in Computer Science*", Publisher: Springer Berlin / Heidelberg. Volume 900/1995, Book"STACS 95", pp. 168-180.

[16] A.A. Samarskii. The Classes of stabilized schemes // *J.comp.math. and math.phys.* (A.A.Samarsky. Classes of stable schemes. Zh. Math. and Math. nat.), 1967, 7, №5. pp. 1096-1133.

In: Several Problems of Applied Mathematics and Mechanics ISBN: 978-1-62081-603-5
Editors: Ivane Gorgidze and Tamar Lominadze

Chapter 6

EXPLORATION OF CURRENT DYNAMICAL PROCESSES IN PNEUMATIC SYSTEMS

***Z. Bogvelishvili*[1], *J. Iosebidze*[1], *G. Abramishvili*[1], *A. Chkheidze*[1], *V. Galashin*[2] *and A. Gustomiasov*[2]**
[1]Georgian Technical University, Tbilisi, Georgia
[2]Bauman Moscow State Technical University, Moscow, Russia

Abstract

A mathematical model has been elaborated to explore pneumatic equalizing spring suspension dynamics, which represents a system of differential equations and describes current processes of thermodynamical, thermal and mass transmission in the research of pneumatic systems.

Keywords: pneumatic systems, air spring suspension

Introduction

Constructive and technological factors have an influence on transport cars pneumatic springs loading characteristics and the air spring suspension working conditions. Exploitation regime of suspension work elasticity elements leads to an important change of working gas. Analytical representation for the dynamic loading characteristics given in the first and second laws of thermodynamics and on the base of transmission of equation make it possible to take into consideration the influence of these factors [1].

Basic Part

Peculiarities of current thermodynamical processes taking place in pneumatic springs with suspension elements in contrast to the usual pneumatic spring processes, make us consider: 1) gas mass variable; 2) energy, which leads to the pneumatic springs together with gas or rejected from them. Besides transport cars, pneumatic springs improvement is connected with the growth of pressing level and static gas pressure in pneumatic springs. That is why at the wide range of pressure and temperature changes, heat exchange with the atmosphere has a considerable influence on the system dynamic. For the description of current thermodynamical processes in transport cars' equalizing pneumatic systems, the calculation energetic method based on thermodynamic equations of mass variables and technical thermodynamics is used to consider the influence of these factors [2,3].

While giving mathematical descriptions of current dynamical processes in the researched air spring suspension, the following fundamental assumptions are made: the working body is an ideal gas; undulatory processes have no place in connecting tubes; the mass of gas in connecting tubes is smaller than the mass of gas in pneumatic spring volume. It is possible to neglect it. Heat exchange in the connecting tube is too little.

To explore pneumatic equalizing air spring suspension dynamics, which represent systems of differential equations and describes thermodynamical heat and mass exchange processes in researched pneumatic system, this mathematical model was worked out. Analytical solutions of these equations in quadratures were received. As pressing and broadening processes pass in different ways, depending on pneumatic spring loading conditions and directions of air stream between them, two possible cases are considered:

1) $p_1 > p_2$ - corresponding to air direction from pneumatic spring 1 into pneumatic spring 2

$$p_1 = p_{01}\varepsilon_1^k \left(\frac{G_1}{G_{01}}\right)^k \exp(-S_1)\left[1 + A\int_0^t \frac{\alpha_1 H_1 V_1^{k-1}}{G_1^k}\exp(S_1)dt\right]; \quad (1)$$

$$p_2 = p_{02}\varepsilon_2^k \exp(-S_2)\left\{1 + B\int_0^t \alpha_2 H_2 V_2^{k-1}\exp(S_2)dt + \int_0^t \left(\frac{V_2}{V_1}\right)^{k-1} G_1^{k-1}\exp(S_2 - S_1)dG_2\left[D + kB\int_0^t \frac{\alpha_1 H_1 V_1^{k-1}}{G_1^k}\exp(S_1)dt\right]\right\}. \quad (2)$$

2) $p_1 < p_2$ - corresponding to air direction from pneumatic spring 2 into pneumatic spring 1

$$p_1 = p_{01}\varepsilon_1^k \exp(-S_1)\left\{1 + E\int_0^t \alpha_1 H_1 V_1^{k-1}\exp(S_1)dt + \int_0^t \left(\frac{V_1}{V_2}\right)^{k-1} G_2^{k-1}\exp(S_1 - S_2)dG_1\left[F + kE\int_0^t \frac{\alpha_2 H_2 V_2^{k-1}}{G_2^k}\exp(S_2)dt\right]\right\}; \quad (3)$$

$$p_2 = p_{02}\varepsilon_2^k \left(\frac{G_2}{G_{02}}\right)^k \exp(-S_2)\left[1 + L\int_0^t \frac{\alpha_2 H_2 V_2^{k-1}}{G_2^k}\exp(S_2)dt\right]. \quad (4)$$

Variable quantities with index "1" belong to the first pneumatic spring, with index "2" to the second pneumatic spring, index "0" shows, that parameter belongs to pressure and broadening start. p, V, T, G and ε are accordingly gas pressure, volume, temperature, mass and level of pressing;

α - coefficient of thermal transmission; R - gas constant; H - surface area of pneumatic spring transmission; T' - temperature of the atmosphere; k - adiabatic exponent; t – time;

$$S_1 = \frac{k-1}{R}\int_0^t \frac{\alpha_1 H_1}{G_1} dt\,;\; S_2 = \frac{k-1}{R}\int_0^t \frac{\alpha_2 H_2}{G_2} dt\,;\; A = \frac{(k-1)T'}{p_{01}}\left(\frac{G_{01}}{V_{01}}\right)^k;$$

$$B = \frac{(k-1)T'}{p_{02}V_{02}^k};\; D = \frac{kp_{01}}{p_{02}G_{01}^k}\left(\frac{V_{01}}{V_{02}}\right)^k;\; E = \frac{(k-1)T'}{p_{01}V_{01}^k};$$

$$F = \frac{kp_{02}}{p_{01}G_{02}^k}\left(\frac{V_{02}}{V_{01}}\right)^k;\; L = \frac{(k-1)T'}{p_{02}}\left(\frac{G_{02}}{V_{02}}\right)^k.$$

Using the calculation dependences for p_1 and p_2, we can define dynamical loading characteristics. Herewith, effective area changes dependence $F_{e1,2}$ on the transference known. Analytical dependences for temperatures to the T_1 and T_2 pneumatic spring 1 – and to the pneumatic spring 2 are received:

1) when $p_1 > p_2$:

$$T_1 = T_{01}\varepsilon_1^{k-1}\left(\frac{G_1}{G_{01}}\right)^{k-1}\exp(-S_1)\left[1 + A^*\int_0^t \frac{\alpha_1 H_1 V_1^{k-1}}{G_1^k}\exp(S_1)dt\right]; \tag{5}$$

$$T_{01} = T_{02}\varepsilon_2^{k-1}\frac{G_{02}}{G_2}\exp(-S_2)\Bigg\{1 + B^*\int_0^t \alpha_2 H_2 V_2^{k-1}\exp(S_2)dt + \int_0^t\left(\frac{V_2}{V_1}\right)^{k-1} G_1^{k-1}\exp(S_2 - \tag{6}$$

$$-S_1)dG_2\left[D^* + kB^*\int_0^t \frac{\alpha_1 H_1 V_1^{k-1}}{G_1^k}\exp(S_1)dt\right]\Bigg\};$$

where

$$A^* = \frac{(k-1)T'}{RT_{01}}\left(\frac{G_{01}}{V_{01}}\right)^{k-1};\; B^* = \frac{(k-1)T'}{RT_{02}V_{02}^{k-1}G_{02}};\; D^* = \frac{kT_{01}}{T_{02}G_{02}G_{01}^{k-1}}\left(\frac{V_{01}}{V_{02}}\right)^{k-1}.$$

2) when $p_1 < p_2$:

$$T_1 = T_{01}\varepsilon_1^{k-1}\frac{G_{01}}{G_1}\exp(-S_1)\left\{1+E^*\int_0^t \alpha_1 H_1 V_1^{k-1}\exp(S_1)dt+\int_0^t\left(\frac{V_1}{V_2}\right)^{k-1}G_2^{k-1}\exp(S_1-\right.$$
$$\left.-S_2)dG_1\left[F^* + kE^*\int_0^t\frac{\alpha_2 H_2 V_2^{k-1}}{G_2^k}\exp(S_2)dt\right]\right\}; \tag{7}$$

$$T_2 = T_{02}\varepsilon_2^{k-1}\left(\frac{G_2}{G_{02}}\right)^{k-1}\exp(-S_2)\left[1+L^*\int_0^t\frac{\alpha_2 H_2 V_2^{k-1}}{G_2^k}\exp(S_2)dt\right]; \tag{8}$$

where

$$E^* = \frac{(k-1)T'}{RT_{01}V_{01}^{k-1}G_{01}};\ F^* = \frac{kT_{02}}{T_{01}G_{01}G_{02}^{k-1}}\left(\frac{V_{02}}{V_{01}}\right)^{k-1};\ L^* = \frac{(k-1)T'}{RT_{02}}\left(\frac{G_{02}}{V_{02}}\right)^{k-1}.$$

The accepted analytical dependences (1)…(8) give the chance to investigate the work of a pneumatic system with the connected elastic elements. From these equations as special cases, it is possible to get the equations describing work at dynamic loadings, systems of the transport car with pneumatic elastic elements with additional volume (let us designate an index «″») and with individual elastic elements:

$$p = p_0\varepsilon^k\left(\frac{G}{G_0}\right)^k\exp(-S^*)\left[1+B\int_0^t\frac{\alpha H V^{k-1}}{G^k}\exp(S^*)dt\right]; \tag{9}$$

$$p'' = p_0''\exp(-S'')\left\{1+B''\int_0^t\alpha''H''\exp(S'')dt+\int_0^t\left(\frac{G}{V}\right)^{k-1}\exp(S''-S^*)dG''\left[D''+\right.\right.$$
$$\left.\left.+kB''\int_0^t\frac{\alpha H V^{k-1}}{G^k}\exp(S^*)dt\right]\right\}; \tag{10}$$

and

$$p'' = p_0''\left(\frac{G''}{G_0''}\right)^k\exp(-S'')\left[1+B''^*\int_0^t\frac{\alpha''H''}{G''^k}\exp(S'')dt\right]; \tag{11}$$

$$p = p_0\varepsilon^k\exp(-S^*)\left\{1+B^*\int_0^t\alpha H V^{k-1}\exp(S^*)dt+\int_0^t V^{k-1}G''^{k-1}\exp(S^*-\right.$$
$$\left.-S'')dG\left[D^* + kB^*\int_0^t\frac{\alpha''H''}{G''^k}\exp(S'')dt\right]\right\}; \tag{12}$$

where $S^* = \frac{k-1}{R}\int_0^t \frac{\alpha H}{G} dt$; $S'' = \frac{k-1}{R}\int_0^t \frac{\alpha'' H''}{G''} dt$; $B = \frac{(k-1)T'}{p_0}\left(\frac{G_0}{V_0}\right)^k$; $B'' = \frac{(k-1)T'}{p''_0 V''}$;

$D'' = \frac{k p_0}{p''_0 V''}\left(\frac{V_0}{G_0}\right)^k$; $B''^* = \frac{(k-1)T' G''^k_0}{p''_0 V''}$; $B^* = \frac{(k-1)T'}{p_0 V_0^k}$; $D^* = \frac{k V'' p''_0}{p_0 V_0^k G''^k_0}$.

Accordingly for temperatures:

$$T = T_0 \varepsilon^{k-1}\left(\frac{G}{G_0}\right)^{k-1} \exp(-S^*)\left[1 + E\int_0^t \frac{\alpha H V^{k-1}}{G^k} \exp(S^*) dt\right]; \tag{13}$$

$$T'' = T''_0 \frac{G''_0}{G''} \exp(-S'')\left\{1 + E''\int_0^t \alpha'' H'' \exp(S'') dt + \int_0^t \left(\frac{G}{V}\right)^{k-1} \exp(S'' - S^*) dG\left[F'' + kE''\int_0^t \frac{\alpha H V^{k-1}}{G^k} \exp(S^*) dt\right]\right\}; \tag{14}$$

and

$$T = T_0 \varepsilon^{k-1} \frac{G_0}{G} \exp(-S^*)\left\{1 + E^*\int_0^t \alpha H V^{k-1} \exp(S^*) dt + \int_0^t V^{k-1} G''^{k-1} \exp(S^* - S'') dG\left[F^* + kE^*\int_0^t \frac{\alpha'' H''}{G''^k} \exp(S'') dt\right]\right\}; \tag{15}$$

$$T'' = T''_0\left(\frac{G''}{G''_0}\right)^{k-1} \exp(-S'')\left[1 + E''^*\int_0^t \frac{\alpha'' H''}{G''^k} \exp(S'') dt\right]; \tag{16}$$

where $E = \frac{(k-1)T'}{RT_0}\left(\frac{G_0}{V_0}\right)^{k-1}$; $E'' = \frac{(k-1)T'}{RT''_0 G''_0}$; $F'' = \frac{kT_0}{T''_0 G''_0}\left(\frac{V_0}{G_0}\right)^{k-1}$;

$E^* = \frac{(k-1)T'}{RT_0 V_0^{k-1} G_0}$; $F^* = \frac{kT''_0}{T_0 V_0^{k-1} G''^{k-1}_0}$; $E''^* = \frac{(k-1)T' G''^{k-1}_0}{RT''_0}$.

With individual elastic elements:

$$p = p_0 \varepsilon^k \exp(-S)\left[1 + A\int_0^t \alpha H V^{k-1} \exp(S) dt\right]; \tag{17}$$

$$T = T_0 \varepsilon^{k-1} \exp(-S)\left[1 + A^*\int_0^t \alpha H V^{k-1} \exp(S) dt\right]; \tag{18}$$

where $S = \frac{k-1}{RG}\int_0^t \alpha H dt$; $A = \frac{(k-1)T'}{p_0 V_0^k}$; $A^* = \frac{(k-1)T'}{RT_0 V_0^{k-1} G_0}$.

The received analytical dependences give the chance to investigate work at dynamic loadings of vehicles with the connected elastic pneumoelements, with additional volume and with individual pneumatic elastic elements, taking into account the thermodynamic processes of heat and mass transfer.

Conclusion

A mathematical model, which describes current thermodynamical processes in spring pneumosystems, is true for any spring pneumatic system, in spite of its construction and elastic elements. From this model, calculating dependences for air spring suspension with additional capacity and for individual pneumatic spring are received.

The analysis for gas pressure changes in equalizing springs showed that pressure differences in the pneumatic spring depends on the following parameters: static pressure and pressing level, speed of pistons and connecting tube diameter moving. Analytical dependences helped to ascertain that pressure changes while pressing in the pneumatic spring depends less on the pressure changes in the pneumatic spring, in which a broadening process takes place. It makes it possible not to take into account ε value change in the pneumatic spring in which the broadening process takes place.

Pressure differences in pneumatic equalizing springs were determined by means of analytical expressions, which showed that a critical regime of gas stream in air spring suspension is not reached. Maximum pressure differences in air spring suspension slightly differs from the air spring suspension with additional capacity. It makes it possible to considerably simplify experimental exploration for the spring with connecting pneumatic spring. As calculations showed, mistakes do not exceed 5...8%. With gas statistic pressure decrease, the mistake will grow.

References

[1] Andersen, B.W. *The Analysis and Design of Pneumatic Systems.* – New York, Krieger Pub Co., 2001, 302 p.

[2] Bogvelishvili, Z., Iosebidze, J. Exploration of Current Dynamical Processes in Automobile Pneumatic Suspension. – *Information Technologies and Management.* Yerevan, 2001, № 2, pp.74-77.

[3] Galashin, V.A., Bogvelishvili, Z.V. Determining load characteristics of vehicle suspension with interconnected pneumatic elements. – Collection of scientific papers MADI "*Investigation and calculation of all-terrain vehicles*", Moscow, 2003, pp.24-28. (In Russ.).

In: Several Problems of Applied Mathematics and Mechanics ISBN: 978-1-62081-603-5
Editors: Ivane Gorgidze and Tamar Lominadze

Chapter 7

Denjoy-Luzin Systems, Absolute Convergence Systems and Unconditional Convergence in Banach Spaces

Vakhtang Kvaratskhelia [*] ***and Vaja Tarieladze*** [†]
Niko Muskhelishvili Institute of Computational Mathematics
of the Georgian Technical University, Tbilisi, Georgia

Abstract

We call an infinite sequence (ξ_k) of random variables given on a probability space $(\Omega, \mathrm{A}, \mathrm{P})$ a Denjoy-Luzin system if for an infinite sequence (x_k) of scalars the almost sure (a.s.) convergence of the series $\sum_{k=1}^{\infty}|x_k\,\xi_k|$ implies $\sum_{k=1}^{\infty}|x_k| < \infty$. We give a characterization of Denjoy-Luzin systems. Moreover, we show that if (ξ_k) is a Denjoy-Luzin system and (x_k) is an infinite sequence of elements of a Banach space X then a.s. unconditional convergence of the series $\sum_{k=1}^{\infty} x_k\,\xi_k$ implies unconditional convergence of the series $\sum_{k=1}^{\infty} x_k$ in X. We show also that a similar result for systems of absolute convergence is true only for finite dimensional Banach spaces.

Keywords: Denjoy-Luzin systems; absolute convergence; unconditional convergence

2010 Mathematics Subject Classifications: Primary 40A05, 40A30, 46B15

---

[*] E-mail address: v_kvaratskhelia@yahoo.com
[†] E-mail address: vajatarieladze@yahoo.com

1. Introduction

Let us recall the following result obtained by A. Denjoy [3] and N. Luzin [4] (see, also, [5], Theorem VI.1.3): Let A be a measurable subset of $[0,2\pi]$ with the strictly positive Lebesgue measure and (α_k) and (x_k) be infinite sequences of real numbers such that

$$\sum_{k=1}^{\infty} |x_k \cos(kt+\alpha_k)| < \infty \text{ for every } t \in A.$$

Then $\sum_{k=1}^{\infty} |x_k| < \infty$.

The following problem arises naturally: to characterize the sequences of measurable functions (ξ_k) for which the following version of Denjoy-Luzin Theorem is valid: For a sequence of scalars (x_k) from almost everywhere convergence of series $\sum_{k=1}^{\infty} |x_k \xi_k|$ follows convergence of the series $\sum_{k=1}^{\infty} |x_k|$.

To investigate the posed problem let us introduce the following definition.

Definition. *An infinite sequence of random variables* (ξ_k) *defined on a probability space* $(\Omega, \mathrm{A}, \mathrm{P})$ *is called a Denjoy-Luzin system if for an infinite sequence of real numbers* (x_k) *the a.s. convergence of the series* $\sum_{k=1}^{\infty} |x_k \xi_k|$ *implies* $\sum_{k=1}^{\infty} |x_k| < \infty$.

In the context of this problem it is expedient to recall the particular case of a result N.N. Vakhania [6] (see, also, [7, p. 30]): A sequence (ξ_k) of standard Gaussian random variables is a Denjoy-Luzin system.

2. Two Characterizations of Denjoy-Luzin Systems

In [8] it is shown that if a sequence (ξ_k) satisfies the condition

$$\mathrm{P}[|\xi_k| \in (m_1, m_2)] \geq \delta > 0 \text{ for some } m_2 > m_1 > 0,$$

then (ξ_k) is a Denjoy-Luzin system. The following theorem gives a complete characterization of Denjoy-Luzin systems.

Theorem 2.1. *An infinite sequence of random variables* (ξ_k) *defined on a probability space* $(\Omega, \mathrm{A}, \mathrm{P})$ *is a Denjoy-Luzin system if and only if the following condition is satisfied*

$$\exists\, r > 0, \qquad \liminf_{k \in N} \mathrm{P}\big[|\xi_k| \geq r\big] > \delta > 0. \tag{1}$$

Proof. Sufficiency. Fix an infinite sequence of numbers (x_k) for which $s = \sum_{k=1}^{\infty} |x_k \, \xi_k| < \infty$ a.s. Clearly there exists a set $B \subset \Omega$, $\mathrm{P}(B) > 1 - \frac{\delta}{2}$, and a number $M > 0$ such that $s(\omega) < M$ for every $\omega \in B$. We have

$$M\,\mathrm{P}(B) \geq \int_B s(\omega)\mathrm{P}(d\omega) \geq \sum_{k=1}^{\infty} |x_k| \int_B |\xi_k(\omega)| \mathrm{P}(d\omega) \geq \sum_{k=1}^{\infty} |x_k| \int_{B \cap [|\xi_k| \geq r]} |\xi_k(\omega)| \mathrm{P}(d\omega) \geq$$

$$\geq r \sum_{k=1}^{\infty} |x_k| \mathrm{P}\big(B \cap [|\xi_k| \geq r]\big).$$

By (1) we can assume that

$$\mathrm{P}\big([|\xi_k| \geq r]\big) > \delta$$

for any indices $k \geq N_\delta$, where the positive integer N_δ depends only on δ. As for any fixed k from the conditions $\mathrm{P}\big([|\xi_k| \geq r]\big) > \delta$ and $\mathrm{P}(B) > 1 - \frac{\delta}{2}$ follows

$$\mathrm{P}\big(B \cap [|\xi_k| \geq r]\big) > \frac{\delta}{2},$$

then

$$M\,\mathrm{P}(B) \geq r \sum_{k=N_\delta}^{\infty} |x_k| \mathrm{P}\big(B \cap [|\xi_k| \geq r]\big) > r \frac{\delta}{2} \sum_{k=N_\delta}^{\infty} |x_k|,$$

or

$$\sum_{k=N_\delta}^{\infty} |x_k| < \left(\frac{2}{r\delta}\right) M\,\mathrm{P}(B) < \infty,$$

and thus $\sum_{k=1}^{\infty} |x_k| < \infty$.

Necessity. Let (ξ_k) be a Denjoy-Luzin system and suppose that condition (1) is not satisfied, i.e. for every $r > 0$

$$\liminf_{k \in N} \mathrm{P}\left[|\xi_k| \geq r\right] = 0.$$

Then, there exists an infinite sequence of indices (k_n), for which

$$\sum_{n=1}^{\infty} \mathrm{P}\left[|\xi_{k_n}| \geq 2^{-n}\right] < \infty.$$

This relation by the Borel-Cantelli Lemma implies that $\sum_{n=1}^{\infty} |\xi_{k_n}| < \infty$ a.s.

Now choose a sequence (x_k) as follows: $x_k = 1$ if $k = k_n$ and $x_k = 0$ – otherwise. Then, clearly $\sum_{k=1}^{\infty} |x_k \, \xi_k| < \infty$ a.s., but $\sum_{k=1}^{\infty} |x_k| = \infty$, which contradicts the condition that (ξ_k) is a Denjoy-Luzin system.

It is obvious that the condition of Theorem 2.1 is satisfied for a sequence of identically distributed not degenerate at zero random variables. Note that in Theorem 2.1 no independence of random variables is required.

Note that if (ξ_k) is a Denjoy-Luzin system and (x_k) is a sequence of scalars then from the a.s. convergence of the series $\sum_{k=1}^{\infty} x_k \, \xi_k$ may not follow the convergence of the series $\sum_{k=1}^{\infty} x_k$. For example, if (ξ_k) is a sequence of independent standard Gaussian random variables then the series $\sum_{k=1}^{\infty} \frac{1}{k} \xi_k$ converges a.s., but the series $\sum_{k=1}^{\infty} \frac{1}{k}$ diverges.

Now consider the case when (x_k) is an infinite sequence of elements in a Banach space X and let (ξ_k) be a Denjoy-Luzin system. If, by analogy the scalar case, under the convergence of series $\sum_{k=1}^{\infty} x_k \, \xi_k$ we will mean the absolute convergence (i.e. $\sum_{k=1}^{\infty} \|x_k \, \xi_k\| < \infty$ a.s.), then from the definition of a Denjoy-Luzin system it follows the absolute convergence of the series $\sum_{k=1}^{\infty} x_k$. But if we consider the unconditional convergence of series instead of the absolute convergence, then the situation is different. Recall that a series $\sum_{k=1}^{\infty} x_k$ in a Banach space converges unconditionally, if its every permutation is convergent. In finite-dimensional Banach spaces the notions of unconditional and absolute convergence are equivalent, but in infinite-dimensional Banach spaces, according to the famous Dvoretzky-Rogers Theorem (see, for example, [9, Theorem 1.c.2]), unconditional convergence is a weaker requirement than absolute convergence. More precisely, in any infinite-dimensional Banach space there exists an unconditionally convergent series, which does not converge absolutely.

Let (ξ_k) be an infinite sequence of random variables defined on a probability space $(\Omega, \mathrm{A}, \mathrm{P})$ and (x_k) is an infinite sequence of elements in a Banach space X. As in [8], we say that the series $\sum_{k=1}^{\infty} x_k\, \xi_k$ converges a.s. unconditionally if there exists a set $\Omega_0 \subset \Omega$, such that $\mathrm{P}(\Omega_0)=1$, and for every $\omega \in \Omega_0$ the series $\sum_{k=1}^{\infty} x_k\, \xi_k(\omega)$ converges unconditionally in X.

Theorem 2.2. *Let (ξ_k) be a sequence of random variables defined on a probability space $(\Omega, \mathrm{A}, \mathrm{P})$ and X be a non-zero Banach space. The following statements are equivalent.*

(i) *(ξ_k) is a Denjoy-Luzin system.*

(ii) *For an infinite sequence (x_k) of elements in X the a.s. unconditional convergence of the series $\sum_{k=1}^{\infty} x_k\, \xi_k$ implies unconditional convergence of the series $\sum_{k=1}^{\infty} x_k$ in X.*

Proof. $(i) \Rightarrow (ii)$. In view of Theorem 2.1 this implication follows from [10, Theorem 1(i)]. For the readers convenience we present a different proof.

We will use the following criteria of the unconditional convergence (see, for example [11], Section 1.3): A series $\sum_{k=1}^{\infty} x_k$ converges unconditionally in a Banach space X if and only if the following two conditions are fulfilled:

$$a)\ \sum_{k=1}^{\infty} \left|\left\langle x^*, x_k \right\rangle\right| < \infty \text{ for every } x^* \in X^*,$$

$$b)\ \lim_{n\to\infty} \sup_{\|x^*\|\le 1} \sum_{k=n}^{\infty} \left|\left\langle x^*, x_k \right\rangle\right| = 0,$$

where X^* is the topological dual of X.

Since (ξ_k) is a Denjoy-Luzin system, then from the a.s. unconditional convergence of the series $\sum_{k=1}^{\infty} x_k \xi_k$ by definition follows the fulfillment of the condition *a*): $\sum_{k=1}^{\infty} \left|\langle x^*, x_k \rangle\right| < \infty$ for every $x^* \in X^*$. Hence, it remains to prove the fulfillment of the condition *b*).

Let us denote

$$\eta_n = \sup_{\|x^*\| \le 1} \sum_{k=n}^{\infty} \left|\langle x^*, x_k \xi_k \rangle\right|, \qquad n = 1,2,\dots .$$

Since the series $\sum_{k=1}^{\infty} x_k \xi_k$ converges a.s. unconditionally, then by *b*) the sequence (η_n) converges a.s. to zero as $n \to \infty$. Take the positive numbers r and δ as in Theorem 2.1. Since the random variable η_1 is a.s. finite, there exists a set $B \subset \Omega$, $\mathrm{P}(B) > 1 - \frac{\delta}{2}$, and a number $M > 0$ such that $\eta_1(\omega) < M$ for every $\omega \in B$. It is clear that (η_n) is the decreasing sequence and for any positive integer n every η_n is integrable on B. Then by the monotone convergence theorem and Theorem 2.1 we have (the symbol 1_B denotes the indicator function of the set B):

$$0 = \lim_{n\to\infty} \int_B \eta_n(\omega)\mathrm{P}(d\omega) \ge \lim_{n\to\infty} \sup_{\|x^*\|\le 1} \sum_{k=n}^{\infty} \left|\langle x^*, x_k \rangle\right| \int_B |\xi_k(\omega)| 1_{B \cap [|\xi_k| \ge r]}(\omega)\mathrm{P}(d\omega) \ge$$

$$\ge r \lim_{n\to\infty} \sup_{\|x^*\|\le 1} \sum_{k=n}^{\infty} \left|\langle x^*, x_k \rangle\right| \mathrm{P}\left(B \cap [|\xi_k| \ge r]\right) \ge r \frac{\delta}{2} \lim_{n\to\infty} \sup_{\|x^*\|\le 1} \sum_{k=n}^{\infty} \left|\langle x^*, x_k \rangle\right| .$$

Hence, the condition *b*) is also fulfilled and, therefore, the series $\sum_{k=1}^{\infty} x_k$ converges unconditionally.

$(ii) \Rightarrow (i)$. This implication follows from the application of (ii) to a one-dimensional subspace of X and the definition of the Denjoy-Luzin system.

The implication $(i) \Rightarrow (ii)$ of Theorem 2.2 generalizes [8, Theorem 2.1].

3. Systems of Absolute Convergence

In this section we consider a concept in a sense dual to our notion of a Denjoy-Luzin system. A sequence (ξ_k) of random variables is an absolute convergence system if the absolute convergence of series of scalars $\sum_{k=1}^{\infty} x_k$ implies the a.s. absolute convergence of series $\sum_{k=1}^{\infty} x_k \, \xi_k$. E.M. Nikishin in [1] gave a characterization of absolute convergence systems. In [2] R.M. Dudley in fact has shown that a sequence (ξ_k) of independent identically distributed random variables is an absolute convergence system if and only if ξ_1 is integrable. The last result together with Theorem 2.1 implies that a Denjoy-Luzin system may fail to be an absolute convergence system.

The following statement shows that an analogue of Theorem 2.2 for systems of absolute convergence remains true only for finite-dimensional Banach spaces.

Theorem 3.1. *Let X be a Banach space. The following statements are equivalent.*

(i) *X is a finite-dimensional Banach space.*

(ii) *For any unconditionally converging series $\sum_{k=1}^{\infty} x_k$ in X and for any system (ξ_k) of absolute convergence the series $\sum_{k=1}^{\infty} x_k \, \xi_k$ converges a.s. unconditionally.*

(iii) *For any sequence of random variables (ξ_k) for which $\sup_k \mathrm{E}|\xi_k| < \infty$ and for any unconditionally converging series $\sum_{k=1}^{\infty} x_k$ in X the series $\sum_{k=1}^{\infty} x_k \, \xi_k$ converges a.s. unconditionally.*

(iv) *For any sequence of independent identically distributed random variables (ξ_k) for which $\mathrm{E}|\xi_1| < \infty$ and for any unconditionally converging series $\sum_{k=1}^{\infty} x_k$ in X the series $\sum_{k=1}^{\infty} x_k \, \xi_k$ converges a.s.*

For the proof of Theorem 3.1 we need two auxiliary assertions which may have an independent interest.

Proposition 3.2. *Let (ξ_k) be a sequence of identically distributed random variables. The following statements are equivalent.*

(i) $C = \sup_{r>0} r^2 \mathrm{P}[|\xi_1| > r] < \infty$.

(ii) $\lim_{k\to\infty} t_k\, \xi_k = 0$ *a.s. for every sequence of real numbers* (t_k) *with* $\sum_{k=1}^{\infty} t_k^2 < \infty$.

(iii) $\xi_1, \xi_2, \ldots,$ *in addition, are independent random variables and* $\lim_{k\to\infty} t_k\, \xi_k = 0$ *a.s. for every sequence of real numbers* (t_k) *with* $\sum_{k=1}^{\infty} t_k^2 < \infty$.

Proof. $(i) \Rightarrow (ii)$. Fix (t_k), $\sum_{k=1}^{\infty} t_k^2 < \infty$, with $|t_k| > 0,\ k = 1,2,\ldots,$ and $\varepsilon > 0$. We have

$$\mathrm{P}[|t_k\, \xi_k| > \varepsilon] = \mathrm{P}\left[|\xi_1| > \frac{\varepsilon}{t_k}\right] \le \frac{C}{\varepsilon^2} t_k^2,\ k = 1,2,\ldots.$$

Hence

$$\sum_{k=1}^{\infty} \mathrm{P}[|t_k\, \xi_k| > \varepsilon] \le \frac{C}{\varepsilon^2} \sum_{k=1}^{\infty} t_k^2 < \infty$$

and (ii) holds by the direct Borel-Cantelli Lemma.

$(ii) \Rightarrow (iii)$ is clear.

$(iii) \Rightarrow (i)$. Denote $v(r) = \mathrm{P}[|\xi_1| > r],\ r > 0$, and suppose (i) is not true. Then

$$\sup_{r>0} r^2 v(r) = \infty.$$

From the last relation we can get the existence of a sequence (r_n) such that

$$r_n > n^2, \qquad r_n^2 v(r_n) > n,\ n = 1,2,\ldots. \tag{3.1}$$

Moreover, there is a sequence (k_n) of natural numbers such that

$$k_1 = 1, \qquad 1 \le \frac{n^2 (k_{n+1} - k_n)}{r_n^2} \le 2,\ n = 1,2,\ldots. \tag{3.2}$$

Let

$$I_n = [k_n, k_{n+1}[,\ n = 1,2,\ldots,$$

and define a sequence (t_k) as follows:

$$t_k = \frac{1}{r_n} \text{ for } k \in I_n,\ n = 1,2,\ldots. \tag{3.3}$$

By (3.3) and (3.2) we have:

$$\sum_{k=1}^{\infty} t_k^2 = \sum_{n=1}^{\infty} \sum_{k \in I_n} t_k^2 = \sum_{n=1}^{\infty} (k_{n+1} - k_n) \frac{1}{r_n^2} \leq \sum_{n=1}^{\infty} \frac{1}{n^2} < \infty .$$

By (3.1), (3.2) and (3.3) we have also,

$$\sum_{k=1}^{\infty} v(1/t_k) = \sum_{n=1}^{\infty} \sum_{k \in I_n} v(1/t_k) = \sum_{n=1}^{\infty} (k_{n+1} - k_n) v(r_n) \geq \sum_{n=1}^{\infty} (k_{n+1} - k_n) \frac{n}{r_n^2} \geq \sum_{n=1}^{\infty} \frac{1}{n} = \infty .$$

Consequently, we found a sequence (t_k), $\sum_{k=1}^{\infty} t_k^2 < \infty$, for which

$$\sum_{k=1}^{\infty} \mathrm{P}\left[|t_k\, \xi_k| > 1 \right] = \infty .$$

From this, since (ξ_k) is a sequence of independent identically distributed random variables by the converse of Borel-Cantelli Lemma we get that (iii) is not satisfied. A contradiction.

Proposition 3.3. *Let X be an infinite dimensional Banach space and (ξ_k) be a sequence of independent identically distributed random variables. Suppose that for any unconditionally converging series $\sum_{k=1}^{\infty} x_k$ in X the series $\sum_{k=1}^{\infty} x_k\, \xi_k$ converges a.s. Then* $\sup_{r>0} r^2\, \mathrm{P}\left[|\xi_1| > r \right] < \infty$.

Proof. Consider a sequence of real scalars (t_k) for which $\sum_{k=1}^{\infty} t_k^2 < \infty$. By Dvoretzky-Rogers Theorem mentioned above there exists unconditionally converging series $\sum_{k=1}^{\infty} x_k$ in X such that $\|x_k\| = |t_k|$, $k = 1,2,\ldots$. As $\sum_{k=1}^{\infty} x_k$ converges unconditionally then by the assumption the series $\sum_{k=1}^{\infty} x_k \xi_k$ converges a.s. Therefore, $(x_k \xi_k)$ converges a.s. to 0, i.e. $\lim_{k\to\infty}\|x_k \xi_k\| = 0$ a.s., that is $\lim_{k\to\infty}|t_k \xi_k| = 0$ a.s. for every (t_k) for which $\sum_{k=1}^{\infty} t_k^2 < \infty$. By Proposition 3.2 the proof is finished.

Proof of Theorem 3.1. (i) implies (ii) by the definition of a system of absolute convergence (since X is finite-dimensional, every unconditionally converging in X series converges absolutely).

(ii) clearly implies (iii), as the condition $\sup_k \mathrm{E}|\xi_k| < \infty$ ensures that (ξ_k) is an absolute convergence system.

Implication $(iii) \Rightarrow (iv)$ is obvious.

Let us proof the implication $(iv) \Rightarrow (i)$.

On the contrary suppose that X is an infinite dimensional Banach space and consider a sequence of independent identically distributed random variables (ξ_k) for which $\mathrm{E}|\xi_1| < \infty$ and $\mathrm{E}|\xi_1|^p = \infty$ for some $p > 1$. Then by Proposition 3.3 there exists in X unconditionally converging series $\sum_{k=1}^{\infty} x_k$ such that the series $\sum_{k=1}^{\infty} x_k \xi_k$ does not converge a.s., which contradicts (iv) and the proof is complete.

Remark. In connection to Theorem 3.1 let us mention the following result obtained in [10]: Let (ξ_k) be a sequence of independent random variables for which $\sup_k \mathrm{E}|\xi_k| < \infty$ and $\sum_{k=1}^{\infty} x_k$ be a series in a Banach space X. If

(a) the series $\sum_{k=1}^{\infty} x_k$ converges unconditionally in X,

and

(b) $\sum_{k=1}^{\infty} x_k \xi_k$ converges a.s. in X,

then $\sum_{k=1}^{\infty} x_k \xi_k$ *converges a.s. unconditionally in* X. Theorem 3.1 shows that in this result the condition (b) is automatically satisfied only in a finite-dimensional case.

Acknowledgments

We are grateful to Professor N. Vakhania for his useful comments and criticism.

References

[1] Nikishin, E.M. Systems of absolute convergence. *Math. USSR*, Sb. 3(1967), 499-507, 1969; translation from Mat. Sb., n. Ser. 74(116), 544-553, 1967.

[2] Dudley, R.M. Random linear functionals. *Trans. Am. Math. Soc.* 136, 1-24, 1969.

[3] Denjoy, A. *Sur l'absolue convergence des sŭries trigonomŭtriques.* C.R. Acad. Sci., Paris 155, 135-136, 1912.

[4] Lusin, N.N. *Sur l'absolue convergence des sŭries trigonomŭtriques.* C.R. Acad. Sci., Paris 155, 580-582, 1912.

[5] Zygmund, A. *Trigonometric series.* Volume I, Cambridge, 1959, 383 p.

[6] Vakhania, N.N. *Sur les rŭpartitions de probabilitŭs dans les espaces de suites numŭriques.* C. R. Acad. Sci., Paris 260, 1560-1562, 1965.

[7] Vakhania, N.N. *Probability Distributions on Linear Spaces.* North Holland, 1981, 137 p.

[8] Kvaratskhelia, V. Unconditional convergence of random series and the geometry of Banach spaces. *Georgian Math. J.* **7**, No.1, 85-96, 2000.

[9] Lindenstrauss, J., Tzafriri, L. *Classical Banach spaces I. Sequence spaces.* Berlin - Heidelberg - New York: Springer-Verlag, 1977, 190 p.

[10] Kwapien, S., Tarieladze, V. On a.s. unconditional convergence of random series in Banach spaces. *Progr. Probab.* 56, 71-75, 2003.

[11] Pietsch A. *Nukleare Lokalkonvexe Rдume.* Akademie-Verlag, Berlin, 1965, 167 p.

[12] Vakhania, N.N., Tarieladze, V.I., Chobanyan, S.A. *Probability distributions on Banach spaces.* Dordrecht: D. Reidel Publishing Company, 1987, 482 p.

In: Several Problems of Applied Mathematics and Mechanics ISBN: 978-1-62081-603-5
Editors: Ivane Gorgidze and Tamar Lominadze

Chapter 8

The Curves of Strong Parabolic Degeneracy as a Part of Boundaries of Domains of Propagation for Some Non-linear Waves

***Jondo Gvazava*[1,*]*, Giorgi Baghaturia*[2,†]**
***and Marine Menteshashvili*[2,‡]**
[1]Georgian Technical University; Tbilisi, Georgia
[2]Niko Muskhelishvili Institute of Computational Mathematics
of the Georgian Technical University; Tbilisi, Georgia

Abstract

The Cauchy problem with closed support for quasi-linear hyperbolic equation with admissible parabolic degeneracy is investigated. The structure of the domain of definition of the solution is described and sufficient conditions for existence of sub-domains impenetrable for that solution are founded.

Keywords: Quasi-linear hyperbolic equation, parabolic degeneracy, Cauchy problem, closed support

2010 Mathematics Subject Classification: 35L72, 35L15

Among several problems posed for hyperbolic equations, it is a well known Riemannian problem. In the case of this problem we have to find unknown solution by given values of the solution and its oblique derivative on the support. At the same time it is also necessary to fulfill the certain conditions regarding initial support. The class of such problems is wide and

* E-mail address: jgvaza@gmail.com
† E-mail address: Nogela@gmail.com
‡ E-mail address: Menteshashvili_m@mail.ru

the known Cauchy problem also belongs to it. Namely, according to the Cauchy problem we have to find the solution of equation by given initial perturbations:

$$u\big|_{t=0} = \tau\,,\ u_t\big|_{t=0} = \nu\,.$$

After posing such problems it is immediately demanded from the initial support, that every characteristic of the equation must not cross it more than in one point. Also it is demanded from the support, that it must not have the characteristic direction in any point. For above-mentioned Cauchy problem, the initial support is represented by relation $t = 0$. This initial support has to be subjected to the same requirements. The existence of these requirements is caused by certain reasons. These reasons are connected with common properties of solutions of the equations. If we consider the equations along the characteristic manifolds, we'll see that in major cases these properties themselves emerge from the equations. For obviousness let us consider the string equation $u_{xx} - u_{tt} = 0$ given on the plane x, t. In this case, as well as in the case of other linear equations, both of the families of characteristics are completely determined by the principal part of the equation. These one-parameter families are related to the characteristic roots $\lambda_1 = 1$, $\lambda_2 = -1$ and they are expressed by relations $x - t = const$, $x + t = const$. If we consider the string equation along the first of them, we'll obtain the relation $du_x - du_t = 0$ by integration of which we obtain $u_x - u_t = const$. It means that along any characteristic of this family the combination of first derivatives of the solution remains constant. The same fact is also true for the other family of the characteristic curves. In this case the combination $u_x + u_t$ remains constant. These combinations are called characteristic invariants. Now, let us assume that the initial support $t = f(x)$ of the Cauchy or the Riemann problem is crossed by the characteristic curve of the first family in two points: (x_0, t_0) and (x_1, t_1). In this case, according to the properties of the characteristic invariants the following relation is fulfilled

$$u_x(x_0,t_0) - u_t(x_0,t_0) = u_x(x_1,t_1) - u_t(x_1,t_1)\,.$$

Here we can make only one conclusion: it turns out that the values of derivatives of the solution are somehow bounded and consequently when posing the problem we cannot take initial arbitrary data on the support. In this case the data are excessive for the correctness of the problem. So, the existence of the above-mentioned condition, posed for the initial support and initial data, is quite natural. However, according to Hadamard [1], the Cauchy problem may appear correct in some cases of closed initial support. Such problems for linear equations were studied in [2-6].

All above mentioned facts are true in case of linear equations, because the characteristic families are determined by the principal part of equation. The situation is quite different when characteristic families cannot be determined in advance, because they depend on an unknown solution u and its derivatives u_x, u_t. This happens in the case of quasi-linear hyperbolic equation

$$au_{xx} + bu_{xt} + cu_{tt} = f\,. \tag{1}$$

The main coefficients and right-hand part of this equation depend on five x, t, u, u_x, u_t variables. In spite of the fact that in such cases the investigation of the problem gets complicated here at least we are freer when posing the problem. We also have the opportunity to modify and generalize the well known linear problems [7], [8]. However, there is no need to generalize and modify the Cauchy problem because it always means to find the solution by given values of it and its non-tangent direction derivative on the support. We emphasize this fact because in the present paper we are going to speak about the Cauchy problem posed for (1) non-linear equation of special type. As it is well known [9], the hyperbolic class of solutions is determined through its characteristic roots by inequality $\lambda_1 \neq \lambda_2$. The characteristic directions also have to be determined in every point by the following relations: $\dfrac{dt}{dx} = \lambda_1(x, t, u, u_x, u_t)$, $\dfrac{dt}{dx} = \lambda_2(x, t, u, u_x, u_t)$. Let us assume that the initial support is given in explicit form by equation $t = f(x)$, where f function is subjected to certain conditions. We'll speak about these conditions below. In this case we can determine the direction of any characteristic curve which comes out from an arbitrary point of the support. This direction is described by the following relation:

$$\frac{dt}{dx} = \lambda_1\left(\tau(x_0), \frac{\tau'(x_0) - \nu(x_0) f'(x_0)}{1 + \left(f'(x_0)\right)^2}, \frac{\nu(x_0) + f'(x_0)\tau'(x_0)}{1 + \left(f'(x_0)\right)^2}\right).$$

If we consider x_0 as a parameter, we can determine the direction of the characteristic curves everywhere on the support. This is an analogous situation we have in the case of the other family of characteristics. However, that is not sufficient, because for the solution of the problem we have to write the equations of the characteristic curves of both families. For this purpose we can use the representations of general integrals. Since the construction of the general integral in closed form is not always possible, here we consider these special type of equations, admitting the construction of General Integrals for them [10]. By means of general integral we have to describe both families of characteristics and construct the set of intersection points of different families of characteristic curves. Generally, such sets create the domain inside of which the initial support is situated. Combinations which remain constant in case of the string equation differ from those for non-linear equations. The main difference is that for non-linear equations these combinations depend on an unknown solution and its first derivatives. In the case of aerodynamics these combinations are called Riemannian invariants. Generally, to construct the characteristic invariants and to use them in solving problems for quasi-linear equations is very difficult. We'll try to do it in a case of a concrete equation, which somehow is related with the flow of air and liquid in channels [11]:

$$u_t(u_t - 1)u_{xx} + (u_t - u_x - 2u_x u_t + 1)u_{xt} + u_x(u_x + 1)u_{tt} = 0. \tag{2}$$

This is a hyperbolic equation; however in some cases it admits a parabolic degeneracy. This happens when the characteristic roots of equation (2) are equal, i.e.

$$-\frac{u_x+1}{u_t}=\frac{u_x}{1-u_t}.$$

So, the set of hyperbolic solutions is described by relation as follows

$$u_x-u_t+1\neq 0.$$

Characteristic invariants are described by the systems:

$$\begin{cases}\xi_1=u+x,\\ \xi_2=\dfrac{u_x}{u_x-u_t+1},\end{cases}\quad \begin{cases}\eta_1=u-t,\\ \eta_2=\dfrac{u_x+1}{u_x-u_t+1}.\end{cases}$$

By means of characteristic method and Poisson's brackets it is possible to obtain the general integral of the equation (2)

$$f(u+x)+g(u-t)=t, \tag{3}$$

where $f,\ g\in C^2(R)$ are two arbitrary functions.

For equation (2) we studied the Cauchy problem posed on the circle:

$$\gamma:\ (x-a)^2+t^2=\left(\sqrt{2}-a\right)^2,\ a<0.$$

Using polar coordinates

$$x=\rho\cos\alpha,\quad t=\rho\sin\alpha$$

we can write this problem as follows:

$$\begin{cases}u\big|_\gamma=\tau(\alpha),\ \ \alpha\in[0,2\pi],\\ u_\rho\big|_\gamma=\nu(\alpha),\ \ t\in[0,2\pi],\quad \tau,\nu\in C^2[0,2\pi],\end{cases} \tag{4}$$

where the support of initial data is the following circle:

$$\gamma:\ \rho=a\cos\alpha+\sqrt{a^2\cos^2\alpha+2-2\sqrt{2}a},\quad \alpha\in[0,2\pi]. \tag{5}$$

If we subject the General integral (3) to initial conditions (4), we obtain the implicit solution for the problem (2), (4):

$$\int_{T_2(u-t)}^{T_1(u+x)} H(\alpha)d\alpha + (2-a)\sin(T_2(u-t)) = t, \tag{6}$$

where

$$H(\alpha) = \frac{(2-a)\big(((a-1)\sin\alpha - \nu(\alpha))\cos\alpha + \tau'(\alpha)\sin\alpha\big)\big(\tau'(\alpha) + (a-2)\sin\alpha\big)}{((a-2)\nu(\alpha) + \tau'(\alpha))\cos\alpha + \big((2-a)\nu(\alpha) + \tau'(\alpha)\big)\sin\alpha + a - 2}$$

and the functions $\alpha = T_1(z)$, $\alpha = T_2(\omega)$ are inverse functions of the following expressions correspondingly:

$$z = \tau(\alpha) + (2-a)\cos\alpha + a,\ \omega = \tau(\alpha) + (2-a)\sin\alpha\,. \tag{7}$$

On the base of (6) it is easy to describe both families of characteristic curves. They are one-parametric families, which come out from arbitrary points of support (5):

$$t = \int_{T_2(c-t-x)}^{c} H(\alpha)d\alpha + (2-a)\sin(T_2(c-t-x)), \tag{8}$$

$$t = \int_{c}^{T_1(c+t+x)} H(\alpha)d\alpha + (2-a)\sin(T_2(c)). \tag{9}$$

As it is known the discriminant curve for the family (8) exists when the following system is solvable in regard to variables x and t:

$$\begin{cases} \displaystyle\int_{T_2(c-t-x)}^{c} H(\alpha)\,d\alpha + (2-a)\sin(T_2(c-t-x)) = t, \\ H(c) - H(T_1(c-t-x))T_1'(c-t-x) + (2-a)\cos(T_2(c-t-x))T_2'(c-t-x) = 0, \end{cases} \tag{10}$$

where the second equation of the system is obtained by derivation of (8) with respect to parameter c.

Analogously, we can write the similar system in the case of family (9):

$$\begin{cases} \displaystyle\int_{c}^{T_1(c+t+x)} H(\alpha)d\alpha + (2-a)\sin(T_2(c)) = t, \\ H(T_1(c+t+x))T_1'(c+t+x) - H(c) + (2-a)\cos(T_2(c))T_2'(c) = 0. \end{cases} \tag{11}$$

It is evident that both systems (10), (11) include τ and ω functions and their derivatives. They also include T_1 and T_2 functions and their derivatives. It should be mentioned, that according to (7), T_1 and T_2 are inverse functions of the expressions that depend only on τ.

Among multiple variants of concrete initial functions τ and ω, we have found an example for the families when (8) and (9) discriminant curve is the same circle:

$$x^2 + t^2 = 1. \tag{12}$$

The initial conditions (4) in this case are as follows:

$$u\big|_\gamma = u_0 + \left(a\cos\alpha + \sqrt{a^2\cos^2\alpha + 2 - 2\sqrt{2}a}\right)\sin\alpha, \quad \alpha \in [0,2\pi],$$

$$u_\rho\big|_\gamma = \sin\alpha, \quad \alpha \in [0,2\pi],$$

where u_0 is the value of the solution in the arbitrary point of the support, and the characteristic curves are straight lines

$$x\cos\alpha + t\sin\alpha = 1,\ \alpha \in [0,2\pi].$$

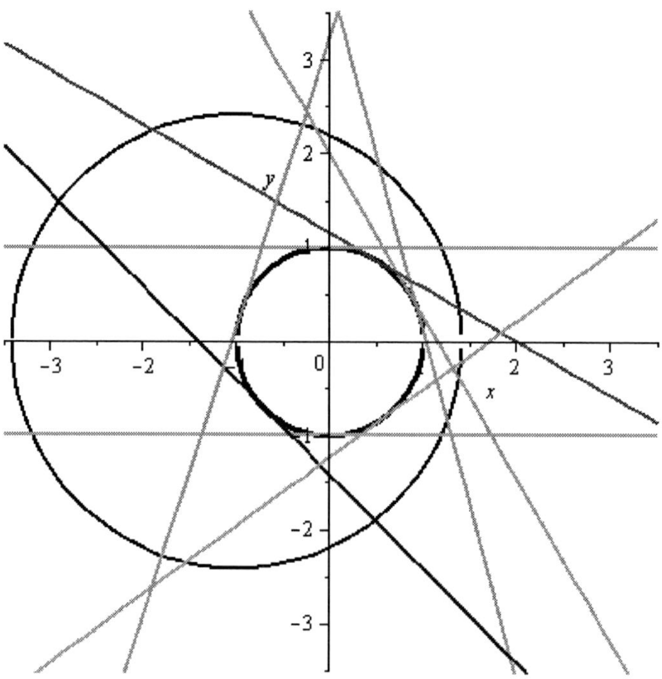

Figure 1.

Each of these straight lines touches the circle (12) at one point (see Fig. 1). This point divides each straight line into two rays. Each of these rays belongs to different families of characteristics. Each line of one family of characteristics crosses the line of other family only one time. At the same time, the circle (12) is the envelope for both families of characteristics. Thus, we have the parabolic degeneracy of the equation on the circle (12). Generally, parabolic degeneracy can be one of the reasons for formation of areas impenetrable for non-linear waves. Such area is the set of inner points of circle (12) in our case.

Thus we have proved the following

Theorem. *The Cauchy problem (2), (4) posed on the closed curve (5) is correct and the solution of the problem is defined everywhere except the inside area of the circle (12).*

References

[1] Hadamard, J. Lecons sur la propagation des ondes et les equations de l'hydrodinamique. *Hermann, Paris*, 1903.

[2] Sobolev, S.L. On mixed problems for partial differential equations with two independent variables. *Dokl. Akad. Nauk SSSR*, 122, 4, 1958, 555-558 (in Russian).

[3] Aleksandryan, R.A. The Boundary Problem of Dirichlet type for degenerative Hyperbolic Equations. *Reports of Symposium in Continuum Mechanics and related problems of Analysis*, Tbilisi, 1971 (in Russian).

[4] Wolfersdorf, L. Zum Problem der Richtungsableitung fur die Tricomi-Gleichung. *Mathematische Nachrichten*, 25, 2, 1963, 111-127.

[5] Vakhania, N.N. On a boundary problem with the prescription on the whole boundary for the hyperbolic system equivalent to the vibrating string equation. *Dokl. Akad. Nauk SSSR*, 116, 1957, 906-909 (in Russian).

[6] Nakhushev, A.M. The criteria of uniqueness of Dirichlet problem for the equation of mixed Type in cylindrical area. *Differ. Uravn.*, 6, 1,1970, 190-191 (in Russian).

[7] Gvazava, J.K. On some classes of Quasi-linear equations of mixed Type. *Metsniereba, Tbilisi*, 1981, (in Russian).

[8] Sokhadze, R.I. First boundary-value problem for an equation of mixed type with weighted compatibility conditions along a parabolic-degeneration line. *Differ. Equations*, 17, 1981, 105-109.

[9] Monge, G. Second memoire sur le calcul integral de quelques equations aux differences partilelles, *Mem.Turin*,annes 1770-1773, 5(2), 1776.

[10] Goursat, E. Lecons sur L'integration des equations aux derives partielles du second ordre. 2. *Hermann, Paris,* 1898.

[11] Webster, A., Sege, G. *Partial Differential Equations of Mathematical Physics*, 1934 (In Russian).

In: Several Problems of Applied Mathematics and Mechanics ISBN: 978-1-62081-603-5
Editors: Ivane Gorgidze and Tamar Lominadze

Chapter 9

ON THE PRICING OF AMERICAN PUT OPTION

***Babilua Petre*[*], *Dochviri Besarion*[†] *and Dochviri Teona*[‡]**
Faculty of Exact and Natural Sciences,
Iv. Javakhishvili Tbilisi State University,
Tbilisi, Georgia

Abstract

A new result is obtained on the vanishing of local time for a non-negative continuous semimartingale at zero. Based on this result, an early exercise premium representation of a value function for the American put option is obtained in a one-dimensional general diffusion model.

2010 Mathematical Subject Classification: 60G48, 60G40, 91B24

Keywords and Phrases: American put option, Snell envelope, local time, value process, optimal stopping

In this paper, we investigate the American put option problem in a general diffusion model and its related optimal stopping problem. The main result of this investigation is the representation of a value function for the American put option as a sum of the corresponding value function of the European type put option and the early exercise premium. In finance literature, this representation is known under the name of the early exercise premium. It was derived by many authors in terms of the well known Black–Scholes model (i.e., when the dynamics of share values is described by geometric Brownian motion). Among them were N. El Karoui and I. Karatzas [5] and S. Jacka [3]. The former authors proceeded from the results in the theory by Balayage of general type random processes, while Jacka used the fact that a value function of the American put option is a solution of the problem with a free boundary (Stefan's problem). A thorough investigation of the American put option problem in the Black–Scholes model was carried out by R. Myneni [8]. It should be noted however that the proofs for many authors somewhat lack clarity.

[*]E-mail address: petre.babilua@tsu.ge
[†]E-mail address: besarion.dochviri@tsu.ge
[‡]E-mail address: dteona@hotmail.com

Our approach is based on studying the properties of local times for general type continuous semimartingales and, as a result, we obtain a sufficient condition (Theorem 3) for the vanishing of local times at zero of non-negative continuous semimartingales. The verification of this condition for the American put option problem has turned out to be a simple matter. In particular, using this condition we derive an early exercise premium representation (Theorem 4) for a value function of the American put option in terms of a general one-dimensional diffusion model. The proof we give is quite clear. Furthermore, our result allows one to directly obtain as a particular case the early exercise premium representation for a value function of the American put option in terms of the well-known Black–Scholes model.

Let $(\Omega, \mathcal{F}, \mathsf{P})$ be a probability space with a standard Wiener process $(W_t, \mathcal{F}_t^W)$, $0 \le t \le T$, defined on it. We will consider on $(\Omega, \mathcal{F}, \mathsf{P})$ the finance market with two assets (B_t, S_t), $0 \le t \le T$, where B_t is the bank account value at time t and S_t is the share price at time t. The evolution of these assets obeys the equations

$$\begin{aligned} dB_t &= r(t) B_t\, dt, \quad 0 \le t \le T, \quad B_0 = 1, \\ dS_t &= r(t) S_t\, dt + \sigma(t, S_t) S_t\, dW_t, \quad 0 \le t \le T, \quad S_0 > 0, \end{aligned} \tag{1}$$

where the continuous functions $r(t)$, $\widetilde{\sigma}(t,x) = \sigma(t,x)x$, $0 \le t \le T$, $x > 0$, are assumed to be such that

$$\begin{aligned} &1)\quad 0 \le r(t) \le R, \\ &2)\quad 0 < \overline{\sigma} \le \sigma(t,x) \le C, \\ &3)\quad \big|\widetilde{\sigma}(t,x) - \widetilde{\sigma}(t,y)\big| \le K|x-y|. \end{aligned} \tag{2}$$

Note that the random process $\frac{S_t}{B_t}$ is a martingale with respect to filtration $\mathcal{F}_t^W$.

Conditions (2) guarantee the existence of a unique strong solution of (1) and also of the following stochastic equation with its initial condition:

$$\begin{aligned} dS_u(t,x) &= r(u) S_u(t,x)\, du \\ &\quad + \sigma\big(u, S_u(t,x)\big) S_u(t,x)\, dW_u, \quad t \le u \le T, \\ S_t(t,x) &= x, \quad x > 0. \end{aligned} \tag{3}$$

Let us consider the American put option for the share S_t with the payoff function $g(x) = (c-x)^+$. The corresponding payoff process $\widetilde{X}_t$ has the form $\widetilde{X}_t = g(S_t)$, while the value process V_t of the American option is written as

$$V_t = \operatorname*{ess\,sup}_{t \le \tau \le T} \mathsf{E}\Big(e^{-\int_t^\tau r(u)\, du} g(S_\tau) \Big|\, \mathcal{F}_t^W \Big), \quad 0 \le t \le T.$$

It is clear that $V_t = B_t Y_t$, where

$$B_t = e^{\int_0^t r(u)\, du}, \quad Y_t = \operatorname*{ess\,sup}_{t \le \tau \le T} \mathsf{E}\Big(e^{-\int_0^\tau r(u)\, du} g(S_\tau) \Big|\, \mathcal{F}_t^W \Big). \tag{4}$$

Note that the value process V_t of the American put option majorizes the payoff process $g(S_t)$, $V_t \ge g(S_t)$, $0 \le t \le T$, and $Y_t \ge X_t$, where

$$X_t = e^{-\int_0^t r(u)\, du} g(S_t), \quad 0 \le t \le T. \tag{5}$$

Here X_t is the discounted payoff process while the process Y_t is its Snell envelope (a minimal supermartingale that majorizes X_t).

Along with the value process of the American put option, we define the value function of this option

$$v^T(t,x) = \sup_{t\le\tau\le T} \mathsf{E}\Big(e^{-\int_t^\tau r(u)\,du}\big(c - S_\tau(t,x)\big)^+\Big), \quad 0\le t\le T, \quad x>0. \tag{6}$$

It is obvious that it is also a value function of the corresponding optimal stopping problem of the nonhomogeneous Markov diffusion process S_t.

Now let us introduce the following notation:

$$D^T \equiv \Big\{(t,x):\ 0\le t<T,\ 0<x,\ v^T(t,x)>g(x)\Big\},$$

where D^T is the continuation domain of stopping problem; D_t^T is the section of D^T at a point t, $0\le t<T$,

$$D_t^T \equiv \{x:\ (t,x)\in D^T\}.$$

We denote by $P(t,x;u,B)$, $u\ge t$, $x>0$, with a Borel subset B of $(0,+\infty)$ the transition probability of the nonhomogeneous diffusion process S_t.

As is known (see [1], p. 445, formula (10)), for the random processes X_t and Y_t the following expansions are valid:

$$X_t = X_0 + m_t + A_t, \quad A_t = A_t^+ - A_t^-, \quad 0\le t\le T, \tag{7}$$

$$Y_t = Y_0 + M_t - C_t, \quad 0\le t\le T, \tag{8}$$

where m_t, M_t are uniformly integrable martingales, while the processes A_t^+, A_t^-, C_t are predictable nondecreasing integrable processes, $A_0^+ = A_0^- = C_0 = 0$.

Theorem 1. *The section of D^T at the point t has the form*

$$D_t^T = \big(b^T(t),+\infty\big), \quad 0\le t<T,$$

where $b^T(t)$ is some function which satisfies the condition $0<b^T(t)<c$, $0\le t<T$.

To derive the early exercise premium representation of the American put option, we need to investigate the properties of a local time for general continuous semimartingales. To this end, let us consider some continuous semimartingale Z_t, $0\le t\le T$, and its set of zeros

$$H = \{(t,w):\ Z_t(w)=0\}. \tag{9}$$

Denote by $L_t^\circ(Z)$ the right local time of the semimartingale at zero. We also need to use the following stopping times

$$D_t(w) = \inf\big\{s>t:\ Z_s(w)\ne 0\big\}\wedge T, \quad 0\le t\le T, \tag{10}$$

and to consider the corresponding predictable random set

$$\widetilde{H}_0 = \bigcup_r (r, D_r], \tag{11}$$

where $(r,D_r]$ denotes the stochastic intervals and the union is taken over all rationals r, such that $0\le r\le T$.

Theorem 2. *Let us consider some continuous semimartingale Z_t, $0 \le t \le T$, and its related set of zeros $H = \{(t,w):\ Z_t(w) = 0\}$. Let us consider its arbitrary, progressively measurable open subset $H_0 \subseteq H$. Then the following relations are valid:*

I. $\int_0^t I_{\widetilde{H}_0}\, dL_s^\circ(Z) = 0$, $\int_0^t I_{\widetilde{H}_0}\, dL_s^\circ(-Z) = 0$,

II. $\int_0^t I_{H_0}\, dL_s^\circ(Z) = 0$, $\int_0^t I_{H_0}\, dL_s^\circ(-Z) = 0$.

In this case, if H_0 is also a predictable random set, then

III. $\int_0^t I_{H_0}\, dZ_s(w) = 0,\ 0 \le t \le T.$

Theorem 3. *Let Z_t be a continuous non-negative semimartingale with its decomposition*

$$Z_t = Z_0 + M_t^z + G_t^+ - G_t^-, \quad 0 \le t \le T, \tag{12}$$

where M_t^z is a local martingale, $G = G_t^+ - G_t^-$ is a process of finite variation, and G_t^+, G_t^- are some nondecreasing processes.

If $\int_0^T I_{(H\setminus \widetilde{H}_0)}\, dG_s^+ = 0$, *then* $L_t^\circ(Z) = 0$ *and* $\int_0^t I_{(Z_s=0)}\, dZ_s = 0$.

Before formulating the main result of this paper on the early exercise premium representation of the American put option, we need to include one more lemma.

Lemma 1. *Consider a random process $X_t(s,x) = \ln S_t(s,x)$, $s \le t \le T$, where $S_t(s,x)$ is the random process defined by equation (3). Then $X_t(s,x)$ satisfies the stochastic differential equation*

$$dX_t(s,x) = \widehat{\mu}\big(t, X_t(s,x)\big)\, dt + \widehat{\sigma}\big(t, X_t(s,x)\big)\, dW_t, \quad s \le t \le T, \tag{13}$$

where

$$\widehat{\mu}(t,x) = r(t) - \frac{1}{2}\sigma^2(t,e^x), \quad \widehat{\sigma}(t,x) = \sigma(t,e^x),$$

and is its unique solution. Moreover, if $c(t)$, $0 \le t \le T$, is some measurable function, then the relations $\mathsf{P}\{X_t(s,x) = c(t)\} = 0$ are valid for all $t \ge s$.

We will now formulate the theorem on representation of the value function of the American put option.

Theorem 4. *Let $b^T(t)$ be the boundary function of the continuation domain defined in Theorem 1. Assume that it has left and right limits $b^T(t+)$, $b^T(t-)$, $0 < t < T$. Then for the value function of the American put option the following early exercise premium representation is valid*

$$v^T(t,x) = \mathsf{E}\Big(e^{-\int_t^T r(u)\,du}\big(c - S_T(t,x)\big)^+\Big)$$
$$+ \int_t^T c r(u) e^{-\int_t^u r(s)\,ds} P(t,x;u,(0,b^T(u)])\, du, \quad 0 \le t \le T, \quad x > 0,$$

where $P(t, x; u, \cdot)$ *is the transition probability for the nonhomogeneous Markov diffusion process* $S_t(s, x)$.

References

[1] P. Babilua, Structure of an American type option value process in a general one-dimensional diffusion model. *Bull. Georgian Acad. Sci.* 170(2004), 443–446.

[2] C. Dellacherie and P.-A. Meyer, Probabilities and Potential. B. Theory of Martingales. *North-Holland Mathematics Studies*, 72, North-Holland Publishing Co., Amsterdam, 1982.

[3] S. D. Jacka, Optimal stopping and the American put. *Math. Finance* 1(1991), No. 2, 1–14.

[4] S. D. Jacka, Local times, optimal stopping and semimartingales. *Ann. Probab.* 21(1993), No. 1, 329–339.

[5] N. El Karoui and I. Karatzas, A new approach to the Skorohod problem, and its applications. *Stochastics Stochastics Rep.* 34(1991), No. 1–2, 57–82.

[6] N. V. Krylov, *Controlled processes of diffusion type.* (Russian) Nauka, Moscow, 1977.

[7] R. S. Liptser and A. N. Shiryaev, Statistics of random processes. I. General theory. (Russian) *Nauka, Moscow,* 1974; English transl.: *Applications of Mathematics (New York),* 5. Stochastic Modelling and Applied Probability. Springer-Verlag, Berlin, 2001.

[8] R. Myneni, The pricing of the American option. *Ann. Appl. Probab.* 2(1992), No. 1, 1–23.

[9] Ph. E. Protter, Stochastic integration and differential equations. 2nd ed. *Applications of Mathematics (New York)*, 21. Stochastic Modelling and Applied Probability. Springer-Verlag, Berlin, 2004.

[10] M. Shashiashvili, Semimartingale inequalities for Snell envelopes. *Stochastics and Stochastics Rep.* 43(1993), No. 1–2, 65–72.

[11] A. N. Shiryaev, Yu. M. Kabanov, D. O. Kramkov, and A. V. Mel'nikov, Toward a theory of pricing options of European and American types. II. Continuous time. (Russian) *Teor. Veroyatnost. i Primenen.* 39(1994), No. 1, 80–129; English transl.: *Theory Probab. Appl.* 39(1994), No. 1, 61–102.

In: Several Problems of Applied Mathematics and Mechanics ISBN: 978-1-62081-603-5
Editors: Ivane Gorgidze and Tamar Lominadze

Chapter 10

ON ONE PROBLEM OF OPTIMAL CONTROL

Babilua Petre * ***and Dochviri Besarion*** †
Faculty of Exact and Natural Sciences,
Iv. Javakhishvili Tbilisi State University,
Tbilisi, Georgia

Abstract

General questions regarding optimal stopping for an inhomogeneous Markov process for time-dependent gain function are investigated. The connection between optimal stopping problems for an inhomogeneous standard Markov process and the corresponding homogeneous Markov process constructed in the extended state space is established. A detailed characterization of a value-function and the limit procedure for its construction in the problem of optimal stopping of an inhomogeneous Markov process is given. The form of ε-optimal (optimal) stopping times is also found.

2010 AMS Subject Classification: 60J05, 91B28, 91B70

Keywords and phrases: Inhomogeneous Markov process, stopping time, payoff, excessive function, extension of state space, universal completion

1.

In this present work, the questions of optimal stopping theory for an inhomogeneous (with infinite lifetime) Markov process for time-dependent gain function and with the observation cost are studied. By extending the state space and the space of elementary events the problems of optimal stopping for the inhomogeneous case can be reduced to the corresponding problems for homogeneous standard Markov processes from which an excessive characterization of a value-function, the method of its construction and the form of ε-optimal (optimal) stopping times for the initial problem are found.

It should be noted that using the method of state space extension in the papers [2], [5], [7], the form of ε-optimal stopping times was established for the case with optimal stopping of homogeneous Markov processes on a bounded time interval.

*E-mail address: petre.babilua@tsu.ge
†E-mail address: besarion.dochviri@tsu.ge

Here, we consider the inhomogeneous (with infinite lifetime) Markov process $X = (\Omega, \mathcal{M}^s, \mathcal{M}_t^s, X_t, P_{s,x})$, $0 \le s \le t < +\infty$, in the state space $(\mathsf{E}, \mathcal{B})$, [1], [3], [6].

Let the gain function $\varphi(t,x)$ and the observation cost $c(t,x) \ge 0$ be Borel measurable functions (i.e., measurable with respect to the product σ-algebra $\mathcal{B}' = \mathcal{B}[0,+\infty) \otimes \mathcal{B}$) which is defined on $\mathsf{E}' = [0,+\infty) \times \mathsf{E}$ and $\varphi(t,x)$ takes its values in $(-\infty, +\infty]$. It is assumed for the observation stopping time t, we obtain a gain

$$g(t,x) = \varphi(t,x) - \int_0^t c(s, X_s)\, ds.$$

It is further assumed that the following integrability condition of a random process $g(t, X_t(\omega)), t \ge 0$, is fulfilled:

$$M_{s,x} \sup_{t \ge s} g^-(t, X_t) < +\infty, \quad s \ge 0, \quad x \in \mathsf{E}. \tag{1}$$

The optimal stopping problem for the process X with gain $g(t,x)$ is stated as follows: the value-function (payoff) $v(s,x)$ is introduced in the form

$$v(s,x) = \sup_{\tau \in \mathfrak{M}_s} M_{s,x} g(\tau, X_\tau), \tag{2}$$

where $\mathfrak{M}_s$ is the class of all finite ($P_{s,x}$-a.s.) M_t^s, $t \ge s$-stopping times; it is required to find the stopping time τ_ε (for each $\varepsilon \ge 0$) for which $M_{s,x} g(\tau_\varepsilon, X_{\tau_\varepsilon}) \ge v(s,x) - \varepsilon$ for any $x \in \mathsf{E}$.

Such a stopping time is called ε-optimal, and in the case $\varepsilon = 0$ it is called simply an optimal stopping time.

To construct ε-optimal (optimal) stopping times, it is necessary to characterize the value $v(s,x)$ and for this purpose the following notion of an excessive function turns out to be fundamental.

2.

Let us introduce now the new space of elementary events $\Omega' = [0,+\infty) \times \Omega$ with elements $\omega' = (s, \omega)$, a new state space (extended state space) $\mathsf{E}' = [0, +\infty) \times \mathsf{E}$ with the σ-algebra $\mathcal{B}' = \mathcal{B}[0,+\infty) \otimes \mathcal{B}$, the new random process X' with values in $(\mathsf{E}', \mathcal{B}')$

$$X_t'(\omega') = X_t'(s,\omega) = (s+t, X_{s+t}(\omega)), \quad s \ge 0, \quad t \ge 0,$$

and the translation operators $\Theta_t'(s,\omega) = (s+t, \omega)$, $s \ge 0, t \ge 0$, where it is obvious that $X_u'(\Theta_t'(\omega')) = X_{u+t}'(\omega')$, $u \ge 0, t \ge 0$.

In the space Ω', we introduce the σ-algebra $N^0 = \sigma(X_u', u \ge 0)$, $N_t^0 = \sigma(X_u', 0 \le u \le t)$ and on the σ-algebra N^0 the probability measures $P_{x'}'(A) = P_{(s,x)}'(A) \equiv P_{s,x}(A_s)$, where $A \in N^0$ and A_s is the section of A at the point s, $A_s = \{\omega : (s,x) \in A\}$, where it is easy to see that $A_s \in \mathcal{F}^s \equiv \sigma(X_u, u = s)$ and if $a \in N_t^0$, then $A_s \in \mathcal{F}_{s+t}^s \equiv \sigma(X_u, s \le u \le s+t)$.

Consider the function $P'(h, x', B') \equiv P'_{x'}(X'_h \in B')$. We have to verify that this function is measurable in x' for a fixed $h \geq 0$. For the rectangles $B' = \Gamma \times B$ which generate the σ-algebra $\mathcal{B}'$ we have

$$P'(h, x', B') = P_{s,x}(\omega : (s+h, X_{s+h}(\omega)) \in \Gamma \times B) = \\ = I_{(s+h \in \Gamma)} P_{s,x}(X_{s+h} \in B).$$

The function $P'(h, x', B')$ is measurable in x', and hence we can introduce the measures $P'_{\mu'}$ on $(\mathsf{E}', \mathcal{B}')$. Let us perform the completion of σ-algebra N^0 with respect to the family of all measures $P'_{\mu'}$, denote this completion by N' and then perform the completion of each σ-algebra N_t^0 in N' with respect to the same family of measures denoting them by N'_t.

Theorem 1. *The random process* $X' = (\Omega', N', N'_t, X'_t, \Theta'_t, P'_{x'})$, $t \geq 0$, *is a homogeneous standard Markov process in the space* $(\mathsf{E}', \mathcal{B}')$.

Let us consider the optimal stopping problem for the process X' with the same gain $g(x') = g(s, x)$ $(x' = (s, x))$ satisfying the conditions

$$M'_{x'} \sup_{t \geq 0} g^-(X'_t) < \infty, \quad P'_{x'}\{\omega' : \lim_{t \downarrow 0} g(X'_t) = g(x')\} = 1, \quad x' \in \mathsf{E}',$$

and with the value $v'(x')$ defined by

$$v'(x') = v'(s, x) = \sup_{\tau' \in \mathfrak{M}'} M'_{x'} g(X'_{\tau'}), \tag{3}$$

where $\mathfrak{M}'$ is the class of all finite ($P'_{x'}$-a.s.) N'_t, $t \geq 0$-stopping times.

Theorem 2. *The values of the initial optimal stopping problem* (3) *coincide* $v(s, x) = v'(s, x)$, $s \geq 0$, $x \in \mathsf{E}$.

The next purpose is an excessive characterization of the payoff $v(s, x)$. Let us note (as can be easily seen) that our definition of a excessive function (with respect to X) coincides exactly with the usual definition of an excessive function (with respect to X'). Therefore, we can directly use Theorem 1, Ch. III in [1] to get the following result.

Theorem 3. *Suppose that condition* (1) *is satisfied. Then the value* $v(s, x)$ *is a minimal excessive majorant of the function* $g(s, x)$. *The value* $v(s, x)$ *is a Borel measurable function* (*i.e.,* $\mathcal{B}'$*-measurable*) *which can be found by the limit procedure*

$$v(s, x) = \lim_{n \to +\infty} \lim_{N \to +\infty} Q_n^N g(s, x),$$

where

$$Q_n g(s, x) = \max\{g(s, x), M_{s,x} g(s + 2^{-n}, X_{2+s^{-n}})\}$$

and Q_n^N *is the* N*-th power of the operator* Q_n.

Theorem 4. *Let the gain* $g(t, x)$ *satisfy* (*with respect to* X) *the following conditions*
1) $M_{s,x} \sup_{t \geq s} |g(t, X_t)| < +\infty$, $s \geq 0$, $x \in \mathsf{E}$;

2) $P_{s,x}\{\omega : \lim_{t\downarrow s} g(t, X_t(\omega)) = g(s, x)\} = 1$, $s \geq 0$, $x \in E$.

Then

i) *for every* $\varepsilon > 0$ *the stopping times*

$$\tau_\varepsilon = \inf\{t \geq s : \; v(t, X_t) \leq g(t, X_t) + \varepsilon\}$$

are ε*-optimal;*

ii) *if the function* $g(t, x)$ *is upper semi-continuous, that is*

$$g(s, x) \geq \varlimsup_{\substack{t\to s \\ y\to x}} g(t, y)$$

and the stopping time $\tau_0(\omega) = \inf\{t \geq s : \; v(t, X_t) = g(t, X_t)\}$ *is finite* ($P_{s,x}$*-a.s.*)*, then* $\tau_0(\omega)$ *is an optimal stopping time.*

References

[1] A. Shiryayev, *Optimal Stopping Rules*. Springer-Verlag, 1978.

[2] B. Dochviri and M. Shashiashvili, On optimal stopping of a homogeneous Markov process on a finite time interval. (Russian) *Math. Nachr.* 156(1992), 269–281.

[3] R. Blumental and R. Getoor, *Markov processes and potential theory*. Academic Press, New York, 1968.

[4] C. Dellacherie and P. Meyer, Probabilities and potential. *North-Holland Math. Studies*, 29, Amsterdam, 1978.

[5] B. Dochviri, Optimal stopping of a nonterminating homogeneous standard Markov process on a finite time interval. *Proc. Steklov Inst. Math.*, 1994, Issue 4, 97–106.

[6] B. Dochviri, On optimal stopping of an inhomogeneous standard Markov process. *Georgian Math. J.* 2(1995), No .4, 335–346.

[7] P. Babilua, I. Bokuchava, B. Dochviri, and M. Shashiashvili, Optimal stopping problem on a finite time interval. *Bull. Georgian Acad. Sci.* 173(2006), No. 2, 250–253.

In: Several Problems of Applied Mathematics and Mechanics ISBN: 978-1-62081-603-5
Editors: Ivane Gorgidze and Tamar Lominadze

Chapter 11

ON ONE MODEL OF REDUCTION OF THE DIRICHLET GENERALIZED PROBLEM TO ORDINARY PROBLEM FOR HARMONIC FUNCTION

Mamuli Zakradze**, *Zaza Sanikidze*†, *Nana Koblishvili*, *Zubiko Natsvlishvili and Niko Muskhelishvili
Institute of Computational
Mathematics of the Georgian Technical University,
Tbilisi, Georgia

Abstract

This paper focuses on the actual practical information regarding boundary problems with the boundary singularities. Dirichlet generalized boundary problem for a Laplace equation is considered for both finite and infinite domains. The case of a boundary function with a limited number of first kind break points is considered under this generalized boundary problem. In this paper, one method is a given ordinary problem for the reduction of the Dirichlet generalized boundary problem with an harmonic function. The method is constructed on the basis of fictitious sources and is applied to finite and infinite domains. From our point of view, this method is characterized by simplicity, stability, and allows for high accuracy. Also, it is oriented for a wide range of users (especially for the researchers in engineering problems). Examples with cusp points in the role of break points are also considered. The results of numerical experiments are given.

Keywords: Dirichlet generalized problem; break point; smoothing; reduction

2010 Mathematics Subject Classifications: 65N38, 65N99, 65E05, 30C30

*E-mail address: mamuliz@yahoo.com
†E-mail address: z_sanikidze@yahoo.com

1. Statement of Generalized Problem and the General Method for Its Reduction to an Ordinary Problem

Let a domain D in the plane $z = x + iy \equiv (x, y)$ be bounded by a piece-wise smooth contour S without multiple points (i.e., S is a simple contour).

It is known that the classical statement of the Dirichlet ordinary boundary value problem for the Laplace equation requires the continuity of the boundary function. However, in practical problems there are cases when the boundary function is piece-wise continuous and therefore it is necessary to consider the Dirichlet generalized problem (see [7, 11]).

A. *On the boundary S of the domain D a function $g(\tau)$ is given which is continuous everywhere, except a finite number of points $\tau_1, \tau_2, \ldots \tau_n$ at which it has discontinuities of the first kind. It is required to find a function $u(z) \equiv u(x, y) \in C^2(D) \bigcap C(\overline{D} \setminus \{\tau_1, \tau_2, \cdots, \tau_n\})$ that satisfy these conditions*

$$\Delta u(z) = 0, \qquad z \in D \tag{1.1}$$

$$u(\tau) = g(\tau), \qquad \tau \in S, \quad \tau \neq \tau_k \ (k = 1, 2, \ldots, n), \tag{1.2}$$

$$|u(z)| < M, \qquad z \in \overline{D}, \tag{1.3}$$

where Δ is the Laplace operator and M is a real constant.

It is known [7, 11] that Problem (1.1)–(1.3) is correct and for the generalized solution $u(z)$ the generalized extremum principle is valid:

$$\min_{z \in S} u(z) < \underset{z \in D}{u(z)} < \max_{z \in S} u(z),$$

where for $z \in S$ it is assumed that $z \neq \tau_k$ $(k = \overline{1, n})$.

Note that the additional requirement of boundedness, when the domain D is finite, concerns actually only the neighborhoods of break points of the function $g(\tau)$. If the domain D is infinite, then condition (1.3) (except as previously-mentioned) means that (see [12])

$$\lim u(z) = c, \quad \text{for } z \to \infty, \tag{1.4}$$

where c is a real constant and $|c| < \infty$. Evidently, if the function $g(\tau)$ is continuous on S, then the Dirichlet generalized problem coincides with the ordinary problem. Condition (1.3) plays an important role in the extremum principle and, consequently, in the theorem on the uniqueness of a solution of Problem **A**.

It should be noted that the methods used for the solution of the Dirichlet ordinary boundary value problem are poorly suited (or not suitable at all) for the solution of the Dirichlet generalized boundary value problem [1–5, 8–10]. In particular, the convergence is very slow and the accuracy is very low in the neighborhood of the singularity for the boundary function. Therefore, researchers try to conduct preliminary improvement of the posed boundary problem. More precisely, they try to reduce, if possible, Problem **A** by smoothing the boundary function $g(\tau)$ to the solution of the ordinary problem [1, 3, 4, 6]. To this end, it is sufficient to have a function $u_0(z)$ which would be a solution of equation (1.1), bounded in $\overline{D}$, continuous in $\overline{D}$ everywhere, except the points $\tau = \tau_k$, and would have the same jumps at the points τ_k, as $g(\tau)$ has. Indeed, if such a function is constructed, then by introduction gives a new unknown function

$$v(z) = u(z) - u_0(z), \tag{1.5}$$

for its determination we have already a Dirichlet ordinary problem.

B.

$$\Delta v(z) = 0, \quad z \in D, \tag{1.6}$$

$$v(\tau) = f(\tau), \quad \tau \in S, \tag{1.7}$$

where $f(\tau)$ is a continuous function on the contour S (since the function $f(\tau)$ at the points τ_k would have removable discontinuities).

If the domain D is infinite, then for the uniqueness of the solution of Problems **B** and **A** ([11, 12]) we require additionally that

$$\lim v(z) = c_1, \quad \text{for } z \to \infty, \tag{1.8}$$

$$\lim u_0(z) = c_2, \quad \text{for } z \to \infty. \tag{1.9}$$

It is evident that in this case, since $c = c_1 + c_2$, c_2 must be given in advance, and c_1 should be found while solving Problem (1.6),(1.7). Conditions (1.4), (1.8), and (1.9) are essential, respectively, for the uniqueness of the solution of Problems **A** and **B** in the case of an infinite domain.

After $v(z) = v(x, y)$ is constructed, (1.5) gives the needed solution $u(z)$,

$$u(z) = v(z) + u_0(z), \quad z \in \overline{D}, \quad z \neq \tau_k. \tag{1.10}$$

If $g^-(\tau_k)$ and $g^+(\tau_k)$ are the limit values of the boundary function $g(\tau)$, when τ tends to the point τ_k along S, respectively, in the positive and negative directions, then the following theorem is valid([6, 7]).

Theorem 1.1. *The limit values of the solution $u(z)$ of the Dirichlet generalized problem, when the point $z \in D$ approaches the point τ_k lie between $g^-(\tau_k)$ and $g^+(\tau_k)$.*

The analysis of the methods considered in [6] (for reduction of the problem **A** to the problem **B**) shows that the following is true:

1. The function $u_0(z)$ which is used for reduction of the Problem **A** to the Problem **B** has various analytical form for when the finite is simply connected, finite multiply connected and infinite domain.

2. For all the break points τ_k there exists the circular neighborhood $G_k(\tau_k, \varepsilon) = \{z : |z - \tau_k| \leq \varepsilon\}$ with the radius ε in the interior of which we may not enter under numerical realization. This fact was expected, since the functions $g(\tau)$, $u(\tau)$ and $u_0(\tau)$ are not defined at the point τ_k.

Thus, under numerical realization of these mentioned methods, actually the Problems **B** and **A** are solved for the domain $D_\varepsilon = D \backslash \{\bigcup_{k=1}^{n} G_k\}$. For the examples, considered in [6], in spite of the fact that calculations were performed with double precision in the role of ε, it turned out the value 10^{-7}. This means that the schemes considered in [6], in general case, do not allow to approach to the break point more closely than 10^{-7}. In other words, they become "unstable" at $\varepsilon < 10^{-7}$.

Evidently, because of correctness of the Problems **A** and **B** such accuracy in approaching the break points is fully sufficient for solution of many practical problems of type **A**. The achievement of more accuracy (e.g., in ballistics, in cosmonautics, in atomic energetics, in physics of high energies, and etc.) schemes of higher accuracy should be constructed.

Here, we propose a more stable and simple scheme(model), which practically gives the possibility to approach the break points with the accuracy allowed by PC.

2. The Model of Reduction of the Dirichlet Generalized Boundary Problem to Ordinary Problem

Let a domain D (finite or infinite) be in the plane $z = x+iy \equiv (x, y)$ with the boundary $S = \bigcup_{k=1}^{m} S_k$, where each S_k is a closed simple contour. Moreover, we assume:

1) The parametric equations of the contours $S_k : x = x_k(t), y = y_k(t), \alpha_k \leq t \leq \beta_k$ is given and $S_k \bigcap S_j = \emptyset$, when $k \neq j$;

2) Under the positive direction the movement along the boundary in the counter-clockwise direction is meant (although, this has no principal sense).

It is evident that in Problem **A** it is not necessary that the break points were placed on all the contours S_k $(k = 1, 2, \ldots, m)$. For simplicity for our writing, we introduce the following notations.

We denote by $\Gamma_1, \Gamma_2, \cdots, \Gamma_l$, where $1 \leq l \leq m$ those from the contours S_k $(k = 1, 2, \ldots, m)$ on which the break points lie, and suppose that the number of the break points on the contour Γ_j is k_j. It is clear that for the natural numbers k_j, we have $1 \leq k_j \leq n$ and $k_1 + k_2 + \cdots + k_l = n$. Further, we denote by $\tau_{j,k}$ $(j = 1, 2, \cdots, l;\ k = 1, 2, \cdots, k_j)$ the break points which lie on the contour Γ_j.

Consider the circles $C_{j,k} = \{z : |z - \tau_{j,k}| = \varepsilon\}$ with the radius ε and with centers at the point $\tau_{j,k}$, respectively. We can take $\varepsilon > 0$ so small that each circle $C_{j,k}$ will intersect the contour Γ_j only at two points $\tau^+_{j,k}$ and $\tau^-_{j,k}$, which are situated, respectively, to the right and to the left of the point $\tau_{j,k}$. Outside the domain D, on the circle $C_{j,k}$ we take the fixed point $\tau^\varepsilon_{j,k}$ in the following way

$$\tau^\varepsilon_{j,k} = (\tau^-_{j,k} - \tau_{j,k})e^{i\alpha_{j,k}} + \tau_{j,k}, \tag{2.1}$$

where i is the imaginary unit.

It is easy to see that if $\alpha_{j,k} > 0$, that then takes place at the rotation vector $\overrightarrow{\tau_{j,k}\tau^-_{j,k}}$; in the positive direction at an angle $\alpha_{j,k}$ about the point $\tau_{j,k}$, and if $\alpha_{j,k} < 0$ then– conversely (it is evident, that in the role of the point $\tau^-_{j,k}$ we can take the point $\tau^+_{j,k}$).

Consider the function $w_{j,k}(z) = \ln\left|\dfrac{z - \tau_{j,k}}{z - \tau^\varepsilon_{j,k}}\right|$, where the point $\tau^\varepsilon_{j,k}$ is defined by the formula (2.1). Evidently, the function $w_{j,k}(z)$ is harmonic in the domain D, continuous in $\overline{D}$ everywhere, except the point $\tau = \tau_{j,k}$ and $\lim_{z\to\infty} w_{j,k}(z) = w_{j,k}(\infty) = 0$. We now show that the function $w_{j,k}(z)$ is bounded in $\overline{D}_\varepsilon$ when $|z-\tau_{j,k}| = \varepsilon$ and $\varepsilon \to 0$, where $D_\varepsilon = D\backslash\{\sum_{j=1}^{l}\sum_{k=1}^{k_j} G_{j,k}\}$ $(G_{j,k}(\tau_{j,k},\varepsilon) = \{z : |z-\tau_{j,k}| \leq \varepsilon\})$. Indeed, suppose that $z \in \overline{D}$ and $|z-\tau_{j,k}| = \varepsilon$. In this case, we have

$$|z-\tau_{j,k}| = \varepsilon,\ |z-\tau^{\varepsilon}_{j,k}| = 2\varepsilon\sin\frac{\theta}{2},\ 0<\theta\leq\pi,$$

and

$$\lim_{z\to\tau_{j,k}} w_{j,k}(z) = -\ln(2\sin\frac{\theta}{2}), \tag{2.2}$$

where θ is the least angle, between the vectors $\overrightarrow{\tau_{j,k}z}$ and $\overrightarrow{\tau_{j,k}\tau^{\varepsilon}_{j,k}}$, i.e., actually the function $w_{j,k}(z)$ is bounded in $\overline{D}$.

It is evident that in the considered case, the circle $C_{j,k}$ contracts to the point $\tau_{j,k}$ $(i.e., C_{j,k} \to \tau_{j,k})$. In particular, from (2.2) we have

$$\varphi^{+}_{j,k} \equiv w^{+}_{j,k}(\tau_{j,k}) = \lim_{\tau\to\tau_{j,k}+} w_{j,k}(\tau) = -\ln(2\sin\frac{\theta^{+}_{j,k}}{2}),$$

$$\varphi^{-}_{j,k} \equiv w^{-}_{j,k}(\tau_{j,k}) = \lim_{\tau\to\tau_{j,k}-} w_{j,k}(\tau) = -\ln(2\sin\frac{\theta^{-}_{j,k}}{2}),$$

where $|\tau-\tau_{j,k}| = \varepsilon, \tau\in\Gamma_j$, $\theta^{+}_{j,k}$ is the angle between the vectors $\overrightarrow{\tau_{j,k}\tau^{+}_{j,k}}$ and $\overrightarrow{\tau_{j,k}\tau^{\varepsilon}_{j,k}}$, and $\theta^{-}_{j,k} \equiv \alpha_{j,k}$ is the angle between the vectors $\overrightarrow{\tau_{j,k}\tau^{-}_{j,k}}$ and $\overrightarrow{\tau_{j,k}\tau^{\varepsilon}_{j,k}}$ $(0<\theta^{+}_{j,k}\leq\pi,\ 0<\theta^{-}_{j,k}\leq\pi)$. In eq.(2.1), we can select the angle $\alpha_{j,k}$ from interval $(0,\pi/2)$ (or from $(-\pi/2,0)$) so that $\varphi^{+}_{j,k}-\varphi^{-}_{j,k} \not\approx 0$.

Suppose now that $\varepsilon > 0$ is required accuracy of approximation to break points and consider the function

$$u_{j,k}(z) = \frac{h_{j,k}}{\delta_{j,k}}\ln\left|\frac{z-\tau_{j,k}}{z-\tau^{\varepsilon}_{j,k}}\right|, \tag{2.3}$$

where $\tau^{\varepsilon}_{j,k}$ is already fixed point on the circle $C_{j,k}$ $(\tau^{\varepsilon}_{j,k} \notin C_{j,k}\cap\overline{D})$. In eq.(2.3), $h_{j,k}$ is a jump of the function $g(\tau)$, and $\delta_{j,k}$ is a jump of the function $w_{j,k}(\tau)$, when passing along Γ_j from point $\tau^{-}_{j,k}$ to the point $\tau^{+}_{j,k}$, i.e.,

$$h_{j,k} = g(\tau^{+}_{j,k}) - g(\tau^{-}_{j,k}),\quad \delta_{j,k} = w_{j,k}(\tau^{+}_{j,k}) - w_{j,k}(\tau^{-}_{j,k}). \tag{2.4}$$

The function $u_{j,k}(z)$ has a certain physical sense. Indeed, consider two the fictitious charge points $q_{j,k} = 2\pi\frac{h_{j,k}}{\delta_{j,k}}$ and $p_{j,k} = -2\pi\frac{h_{j,k}}{\delta_{j,k}}$, which lie in z- plane, respectively at the points $\tau_{j,k}$ and $\tau^{\varepsilon}_{j,k}$. Potential field, created with couple of the charges $q_{j,k}$ and $p_{j,k}$, at the point z is

$$u_{j,k}(z) = \frac{h_{j,k}}{\delta_{j,k}}\ln|z-\tau_{j,k}| - \frac{h_{j,k}}{\delta_{j,k}}\ln|z-\tau^{\varepsilon}_{j,k}| = \frac{h_{j,k}}{\delta_{j,k}}\ln\left|\frac{z-\tau_{j,k}}{z-\tau^{\varepsilon}_{j,k}}\right|,$$

where $\ln|z-\tau_{j,k}|$ and $\ln|z-\tau^{\varepsilon}_{j,k}|$ are the fundamental solutions of Laplace equation [7].

Let us denote by $\tau^*_{j,k}$ a "object point" (singular point of field) which is obtained by approach of the charges $q_{j,k}$ and $p_{j,k}$. On the basis of selection for $\varepsilon > 0$, we can take the point $\tau^{\varepsilon}_{j,k}$ so near to the point $\tau_{j,k}$ that "object point" $\tau^*_{j,k}$ should be in the tolerance field of the boundary S; i.e., in the limits, in which boundary S may change practically arbitrarily (without change of type). Actually, under numerical realization, in the role of the point $\tau^*_{j,k}$ we understand the disk $|z-\tau_{j,k}| < \varepsilon$.

Thus, we constructed the function $u_{j,k}(z)$, which is harmonic in the domain D, bounded in $\overline{D}_{\varepsilon}$ (for $\varepsilon \to 0$), $u_{j,k}(\infty) = 0$ and continuous in $\overline{D}$ everywhere, except the points $\tau_{j,k}$.

When passing along Γ_j in the positive direction through the point $\tau^*_{j,k}$, the function $u_{j,k}(z)$ has a jump

$$u_{j,k}(\tau^+_{j,k}) - u_{j,k}(\tau^-_{j,k}) = h_{j,k}.$$

It is evident, for smoothing the function $g(\tau)$ on the contour Γ_j we can use the function

$$u_j(z) = \sum_{k=1}^{k_j} u_{j,k}(z),$$

and in the role $u_0(z)$, we take

$$u_0(z) = \sum_{j=1}^{l} u_j(z) = \sum_{j=1}^{l}\sum_{k=1}^{k_j} u_{j,k}(z).$$

Consider the function

$$v(z) = u(z) - \sum_{j=1}^{l}\sum_{k=1}^{k_j} u_{j,k}(z),$$

where $u(z)$ is a solution of Problem **A** for the domain D with the boundary function $g(\tau)$. We now show that the function $v(z)$ is a solution to the Problem **B** for the domain D.

Indeed, the function $u(z)$ and all the functions $u_{j,k}(z)$ (see (2.3)) are harmonic in the domain D. Further, the limit value of the function $v(z)$, when $z \to \tau$, $(z \in D)$ and $\tau \in S^*$ $(S^* = S \backslash \{\sum_{j=1}^{l}\sum_{k=1}^{k_j} L_{j,k}\}$, $L_{j,k}$ is open arc of the contour Γ_j between the points $\tau^-_{j,k}$ and $\tau^+_{j,k})$ is

$$f(\tau) = g(\tau) - \sum_{j=1}^{l}\sum_{k=1}^{k_j} u_{j,k}(\tau), \tag{2.5}$$

and the function $f(\tau)$ remains continuous while passing through every point $\tau^*_{j,k}$(for $\varepsilon \to 0$). Indeed, in (2.5), from the function $g(\tau)$ (with a jump $h_{j,k}$ at the point $\tau^*_{j,k}$), we subtract the function $u_{j,k}(\tau)$ having the same jump, and the rest of the terms for the sum (2.5) are continuous at this point.

Thus, the approximate solution to Problem **A** for the domain D can be represented in the form

$$u(z) = v(z) + \sum_{j=1}^{l} \sum_{k=1}^{k_j} u_{j,k}(z),$$

where $v(z)$ is a approximate solution to the Problem **B** with the boundary function $f(\tau)$ for the domain D and the functions $u_{j,k}(z)$ are given by (2.3).

Remark 2.1. At calculation of the values $h_{j,k}$, $\delta_{j,k}$ and the function $u_{j,k}(z)$, we cannot infinitely approach the point $\tau_{j,k}$. Indeed, in this case (see 2.3,2.4) at calculation of functions $u_{j,k}(z)$ overflow may occur(i.e., stoppage of a computer). At numerical realization, from our point of view, the following approach is more appropriate. Since the functions $g(\tau)$ and $f(\tau)$ are continuous in the neighborhood for the points $\tau_{j,k}$, therefore, for arbitrarily small $\varepsilon_1 > 0$ there exists such $\varepsilon(\varepsilon_1)$ that the following conditions are fulfilled: 1) each circle $C_{j,k}$ intersects the contour Γ_j only at two points $\tau_{j,k}^+$ and $\tau_{j,k}^-$, which are situated, respectively, to the right and to the left of the point $\tau_{j,k}$; 2) there will be no difficulties in calculations; 3) inequalities $|f(\tau_{j,k}^+) - f(\tau_{j,k}^-)| < \varepsilon_1$ are valid. It is evident that in this case, we may say that the function $f(\tau)$ at the point $\tau_{j,k}$ has a removable singularity with accuracy ε_1,i.e., the function $f(\tau)$ is practically continuous at point $\tau_{j,k}$.

If the parametric equation of the contour Γ_j is $z = z_j(t)$ and $\tau_{j,k} = z_j(t_{j,k})$, then we can take ε so small that in the role of the points $\tau_{j,k}^+$ and $\tau_{j,k}^-$, we can take $\tau_{j,k}^+ = z_j(t_{j,k}+\varepsilon)$ and $\tau_{j,k}^- = z_j(t_{j,k} - \varepsilon)$, and the conditions 1), 2), and 3) will be fulfilled.

3. Numerical Examples for Smoothing the Boundary Functions

In this section, we present the results of the numerical experiments which are performed on the basis of the proposed method. In order to compare the previously mentioned method (in the sense of smoothing accuracy) and methods given in [6] we consider the same examples, as in [6]. In Examples 3.1.–3.4., with domain D, we consider finite simply connected domains and the z-planes with one hole, i.e., m=1 and $S = S_1 = \Gamma_1$. For the sake of simplicity, In Examples 3.1.–3.4., we take as the function $g(\tau)$ and the same function which has four points $\tau_{1,k}(k = 1, 2, 3, 4)$ of finite discontinuity. The points $\tau_{1,k}$ lie on the contour S preserving the order of succession under the positive circuit of S. In particular, we take the function

$$g(\tau) = \begin{cases} 1 & \text{for} \quad \tau \in \tau_{1,1}\tau_{1,2}, \\ 2 & \text{for} \quad \tau \in \tau_{1,2}\tau_{1,3}, \\ 3 & \text{for} \quad \tau \in \tau_{1,3}\tau_{1,4}, \\ 4 & \text{for} \quad \tau \in \tau_{1,4}\tau_{1,1}, \end{cases}$$

where $\tau_{1,1}\tau_{1,2}$, $\tau_{1,2}\tau_{1,3}$, $\tau_{1,3}\tau_{1,4}$, $\tau_{1,4}\tau_{1,1}$ are open arcs of the contour S. It is evident that the jumps of the function $g(\tau)$ at the points $\tau_{1,k}(k = 1, 2, 3, 4)$ are equal: $h_{1,1} = -3$; $h_{1,2} = 1$; $h_{1,3} = 1$; $h_{1,4} = 1$. According to Remark 2.1., as the points $\tau_{1,k}^+$ and $\tau_{1,k}^-$ we take $\tau_{1,k}^+ = z(t_{1,k}+\varepsilon)$, $\tau_{1,k}^- = z(t_{1,k}-\varepsilon)$ respectively, where $z = z(t)$ is the parametric equation of the contour S, and $t_{1,k}$ is the value of parameter t for which $\tau_{1,k} = z(t_{1,k})$. The values of function $f(\tau)$ and of $\delta_{1,k} = w_{1,k}(\tau_{1,k}^+) - w_{1,k}(\tau_{1,k}^-)$ at points $\tau_{1,k}^{\pm}$ for different

ε are given in the tables. In the numerical experiments, calculations were performed in double precision.

Suppose we are interested in such a case, when the function $g(\tau)$ has only one break point τ^* on the contour S. For this, it is sufficient to take $g(\tau)$ in the following form

$$g(\tau) = \begin{cases} \psi_1(\tau)\,, & \tau \in \tau^*\tau^{**}, \\ \psi_2(\tau)\,, & \tau \in \tau^{**}\tau^*, \end{cases}$$

where $\tau^{**} \neq \tau^*$, $\tau^{**} \in S$, the function $\psi_1(\tau)$ and $\psi_2(\tau)$ are continuous on the closed arcs $\tau^*\tau^{**}$ and $\tau^{**}\tau^*$, respectively and $\psi_1(\tau^{**}) = \psi_2(\tau^{**})$, $\psi_1(\tau^*) \neq \psi_2(\tau^*)$.

Example 3.1. The domain D is the interior of the ellipse $S:\ x = a\cos t,\ y = b\sin t,\ 0 \leq t \leq 2\pi$. In Table 3.1, the results are given for $a = 5,\ b = 2$ and $\tau_{1,1} = (5,0),\ \tau_{1,2} = (0,2),\ \tau_{1,3} = (-5,0),\ \tau_{1,4} = (0,-2)$. For finding the points $\tau^{\varepsilon}_{1,k}$ (k=1,2,3,4) in (2.1) we take $\alpha_{1,k} = \pi/3$ (k=1,2,3,4)

Table 3.1.

ε	k	$\delta_{1,k}$	ε_1
	1	-0.549313361110	
	2	-0.549307299038	
10^{-5}	3	-0.549313361132	10^{-10}
	4	-0.549307299050	
	1	-0.549306216865	
	2	-0.549306156669	
10^{-7}	3	-0.549306217884	10^{-11}
	4	-0.549306157565	
	1	-0.549305078472	
	2	-0.549306551117	
10^{-10}	3	-0.549306956814	10^{-12}
	4	-0.549307875299	
	1	-0.536678956760	
	2	-0.555585678350	
10^{-14}	3	-0.565265035557	$10-13$
	4	-0.568619035142	

Example 3.2. As an infinite domain D, we take the exterior of the ellipse $S:\ x = a\cos t,\ y = b\sin t,\ 0 \leq t \leq 2\pi$. In Table 3.2, we represent the results of calculations for $a = 5,\ b = 2,\ ,\ \tau_{1,1} = (5,0),\ \tau_{1,2} = (0,2),\ \tau_{1,3} = (-5,0),\ \tau_{1,4} = (0,-2)$ and $\alpha_{1,k} = -\pi/3$ (k=1,2,3,4).

Example 3.3. The domain D is the interior of the astroid $S:\ x = 2\cos^3 t,\ y = 2\sin^3 t,\ 0 \leq t \leq 2\pi$. In Table 3.3, we present the results for the case when the points $\tau_{1,k}(k = 1,2,3,4)$ are the points of cusp, namely $\tau_{1,1} = (2,0),\ \tau_{1,2} = (0,2),\ \tau_{1,3} =$

Table 3.2.

ε	k	$\delta_{1,k}$	ε_1
	1	-0.549298927352	
	2	-0.549304989637	
10^{-5}	3	-0.549298927374	10^{-9}
	4	-0.549304989649	
	1	-0.549306073284	
	2	-0.549306133337	
10^{-7}	3	-0.549306074303	10^{-12}
	4	-0.549306134746	
	1	-0.549305078472	
	2	-0.549306807512	
10^{-10}	3	-0.549306956814	10^{-13}
	4	-0.549307875299	
	1	-0.536678956760	
	2	-0.555585678350	
10^{-14}	3	-0.565265035557	10^{-13}
	4	-0.568619035142	

$(-2,0)$, $\tau_{1,4} = (0,-2)$. For each ε (in the sense of smoothing accuracy) are selected corresponding angles $\alpha_{1,k}(k = 1,2,3,4)$ (see (2.1).

Table 3.3.

ε	$\alpha_{1,k}$	k	$\varepsilon_1)$
		1	
		2	
10^{-2}	$\pi 10^{-6}$	3	10^{-10}
		4	
		1	
		2	
10^{-5}	$\pi 10^{-8}$	3	10^{-13}
		4	
		1	
		2	
10^{-7}	$\pi 10^{-10}$	3	10^{-13}
		4	

Example 3.4. The domain D is the exterior of the astroid S : $x = 2\cos^3 t$, $y = 2\sin^3 t$, $0 \leq t \leq 2\pi$. In the role for the points $\tau_{1,k}(k = 1,2,3,4)$ are taken $\tau_{1,1} = (2,0)$, $\tau_{1,2} = (0,2)$, $\tau_{1,3} = (-2,0)$, $\tau_{1,4} = (0,-2)$. In the role for points $\tau_{1,k}^{\varepsilon}$ (k=1,2,3,4)

are for the points which divide the segments $\tau_{1,k}^{-}\tau_{1,k}^{+}$ in parts proportional to the values $m_{1,k}$ and $n_{1,k}$ (evidently, we can define location of the points $\tau_{1,k}^{\varepsilon}$ (k=1,2,3,4) with the help of formula (2.1), i.e.,

$$\frac{m_{1,k}}{n_{1,k}} = \frac{|\tau_{1,k}^{+} - \tau_{1,k}^{\varepsilon}|}{|\tau_{1,k}^{-} - \tau_{1,k}^{\varepsilon}|}$$

and

$$\tau_{1,k}^{\varepsilon} = \frac{n_{1,k}\tau_{1,k}^{+} + m_{1,k}\tau_{1,k}^{-}}{m_{1,k} + n_{1,k}}.$$

For each value of ε numbers $m_{1,k}$ and $n_{1,k}$ are selected in the sense of "best" smoothing of boundary function. The results of the calculation are given in the Table 3.4.

Table 3.4.

ε	$m_{1,k}$	$n_{1,k}$	k	ε_1
10^{-2}	3	1	1 2 3 4	10^{-12}
10^{-5}	10	1	1 2 3 4	10^{-13}
10^{-7}	1000	1	1 2 3 4	10^{-13}

Example 3.5. D is the doubly-connected domain with the boundary $S = S_1 \cup S_2$, where the contour $S_1(S_1 \equiv \Gamma_1)$ is the ellipse $S_1 : x = a\cos t,\ y = b\sin t$, while the contour $S_2(S_2 \equiv \Gamma_2)$ is the circumference $S_2 : x = r\cos t,\ y = r\sin t,\ 0 \leq t \leq 2\pi$. As the boundary function, we take the following one:

$$g(\tau) = \begin{cases} g_1(\tau), & \tau \in S_1, \\ g_2(\tau), & \tau \in S_2. \end{cases} \tag{3.1}$$

In (3.1), the functions $g_1(\tau)$ and $g_2(\tau)$ on the open arcs of the contours S_1 and S_2, respectively have the form

$$g_1(\tau) = \begin{cases} 1 & \text{for } \tau \in \tau_{1,1}\tau_{1,2}, \\ 2 & \text{for } \tau \in \tau_{1,2}\tau_{1,3}, \\ 3 & \text{for } \tau \in \tau_{1,3}\tau_{1,4}, \\ 4 & \text{for } \tau \in \tau_{1,4}\tau_{1,1}, \end{cases} \qquad g_2(\tau) = \begin{cases} 1 & \text{for } \tau \in \tau_{2,1}\tau_{2,2}, \\ 3 & \text{for } \tau \in \tau_{2,2}\tau_{2,3}, \\ 5 & \text{for } \tau \in \tau_{2,3}\tau_{2,4}, \\ 7 & \text{for } \tau \in \tau_{2,4}\tau_{2,1}. \end{cases}$$

The jumps of the function $g(\tau)$ at the points of discontinuity $\tau_{j,k}$ $(j = 1, 2; k = 1, 2, 3, 4)$ equal: $h_{1,1} = -3, h_{1,2} = 1, h_{1,3} = 1, h_{1,4} = 1, h_{2,1} = -6, h_{2,2} = 2, h_{2,3} = 2, h_{2,4} = 2$. The results of the calculations for $a = 5, b = 2, r = 10, \tau_{1,1} = (5, 0), \tau_{1,2} = (0, 2), \tau_{1,3} = (-5, 0), \tau_{1,4} = (0, -2), \tau_{2,1} = (10, 0), \tau_{2,2} = (0, 10), \tau_{2,3} = (-10, 0), \tau_{2,4} = (0, -10)$ and for various values of ε are given in Tables 3.5–3.7. In numerical experiments for the points $\tau^{\varepsilon}_{1,k}$, $\alpha_{1,k} = -\pi/3$, and for the points $\tau^{\varepsilon}_{2,k}$, $\alpha_{2,k} = \pi/3$ $(k = 1, 2, 3, 4)$ (see (2.1)).

Table 3.5.

$\varepsilon = 10^{-5}$		
k	$\delta_{1,k}$	ε_1
1	−0.549298927352	
2	−0.549304989637	
3	−0.549298927374	10^{-8}
4	−0.549304989649	
k	$\delta_{2,k}$	ε_1
1	−0.549309031068	
2	−0.549309031070	
3	−0.549309031079	10^{-9}
4	−0.549309031087	

Table 3.6.

$\varepsilon = 10^{-7}$		
k	$\delta_{1,k}$	ε_1
1	−0.549306073284	
2	−0.549306133337	
3	−549306074303	10^{-12}
4	−549306134746	
k	$\delta_{2,k}$	ε_1
1	−0.549306172765	
2	−0.549306173976	
3	−0.549306173784	10^{-13}
4	−0.549306174872	

Finally, it should be noted that the analysis of the above-described model shows the following:

a) The function $u_0(z)$ has one and the same analytic structure for finite and infinite domains;

b) Computer programming for smoothing process of boundary function, and numerical realization are easy and they are not associated with difficulties;

Table 3.7.

$\varepsilon = 10^{-10}$		
k	$\delta_{1,k}$	ε_1
1	-0.549305078472	
2	-0.549306807512	
3	-0.549306956814	10^{-13}
4	-0.549307875299	
k	$\delta_{2,k}$	ε_1
1	-0.549306104051	
2	-0.549307063907	
3	-0.549306956814	10^{-13}
4	-0.549307875299	

c) the considered model is stable with respect to calculation errors, and is distinguished by high accuracy of approach to the break points. In particular, for one and the same accuracy $\varepsilon = 10^{-7}$ of approach to the break points, the methods considered in [6] give the accuracy of smoothing of the boundary function $\varepsilon_1 = 10^{-6}$, and the method proposed by us above, gives $\varepsilon_1 = 10^{-13}$.

On the basis of the noted above we think that the model considered in this paper will be suitable for wide range of users.

References

[1] Aleksidze, M.A. (1991). *Fundamental Functions in Approximate Solutions of the Boundary Value Problems* (in Russian), Part I, Nauka, Moscow.

[2] Fairweather, G., and Karageorghis, A. (1998). The method of fundamental solutions for elliptic boundary value problems. *Adv. Comput. Math.* 9, 69–95.

[3] Grinberg, G.A. (1948). *The Selected Questions of Mathematical Theory of Electric and Magnetic Phenomena* (in Russian), Academic Press, USSR.

[4] Kantorovich, L.V., and Krylov, V. I. (1962). *Approximate Methods of Higher Analysis* (in Russian), 5th ed., Fizmatgiz, Moscow-Leningrad.

[5] Karageorghis, A. (1992). Modified methods of fundamental solutions for harmonic and biharmonic problems with boundary singularities. *Numer. Methods Partial Differential Equations.* 8, 1–19.

[6] Koblishvili, N.,Tabagari, Z., and Zakradze, M. (2003). On reduction of the Dirichlet generalized boundary value problem to an ordinary problem for harmonic function. *Proc. A.Razmadze Math. Inst.* 132, 93–106.

[7] Lavrent'jev, M.A., and Shabat, B. V. (1973). *Methods of the Theory of Functions of a Complex Variable* (in Russian), Nauka, Moscow.

[8] Li Z-C., Mathon, R., and Sermer, P. (1987). Boundary methods for solving elliptic problems with singularities and interfaces. *SIAM J.Numer. Anal.* 24, 487–498.

[9] Olson, L.G., Georgiou, G.C., and Schultz, W.W. (1991). An efficient finite element method for treating singularities in Laplace's equation. *J. Comput. Phys.* 96, 391–409.

[10] Poullikkas, A., Karageorghis, A., and Georgiou, G. (1998). Methods of fundamental solutions for harmonic and biharmonic boundary value problems. *Comput. Mechanics.* 21, 416–423.

[11] Tikhonov, A.N., and Samarski, A. A. (1972). *Equations of Mathematical Physics* (in Russian), Nauka, Moscow.

[12] Vladimirov, V.S. (1971). *Equations of Mathematical Physics* (in Russian), Nauka, Moscow.

In: Several Problems of Applied Mathematics and Mechanics ISBN: 978-1-62081-603-5
Editors: Ivane Gorgidze and Tamar Lominadze

Chapter 12

SPECIAL UNIVERSAL SERIES*

***Giorgi Giorgobiani*† and *Vaja Tarieladze*‡**
Institute of Computational Mathematics of the Georgian
Technical University, Tbilisi, Georgia

Abstract

We show that in an arbitrary separable Banach space X there exists an universal series $\sum_i x_i$ with the following additional properties: $\sum_{i=1}^{\infty} \|x_i\|^p < \infty$ for each $p \in [1, \infty]$; the series $\sum_i \vartheta_i x_i$ converges in X for almost all sequences of signs (ϑ_i). In a wide class of complete separable metrizable topological vector spaces the existence of an universal series with property (2) is established as well.

2000 Mathematics Subject Classification: 46A35, 46B13, 46B15

Keywords and Phrases: Banach space, sum range, almost unconditional convergence, universal series

1. Introduction

Let Y be a real Hausdorff topological vector space and (y_n) be a sequence of elements of Y.

For (y_n) let us associate a subset $SR(\sum_{k=1}^{\infty} y_k)$ of Y as follows: an element $y \in Y$ belongs to $SR(\sum_{k=1}^{\infty} y_k)$ if there exists a permutation $\pi : \mathbb{N} \to \mathbb{N}$ such that the sequence $(\sum_{k=1}^{n} y_{\pi(k)})_{n\in\mathbb{N}}$ converges to y. The set $SR(\sum_{k=1}^{\infty} y_k)$ is called *the sum range* of (y_n) (cf. [8, Definition 2.1.1]).

A series $\sum_{k=1}^{\infty} y_k$ is called *unconditionally convergent* if for every permutation $\pi : \mathbb{N} \to \mathbb{N}$ the series $\sum_{k=1}^{\infty} y_{\pi(k)}$ converges in Y.

It is known that if the series $\sum_{k=1}^{\infty} y_k$ is unconditionally convergent, then $SR(\sum_{k=1}^{\infty} y_k)$ is a singleton.

The following statement is a classical result of Mathematical Analysis.

---

*The paper is partially supported by grant GNSF/ST06/3-009.
†E-mail address: bachanabc@yahoo.com
‡E-mail address: vajatarieladze@yahoo.com

Theorem 1.1. (B. Riemann) *If a series $\sum_n t_n$ consisting of real numbers converges but does not converge unconditionally, then $SR(\sum_{k=1}^{\infty} t_k) = \mathbb{R}$.*

For a finite-dimensional Y and a convergent but not unconditional convergent series $\sum_n y_n$ its sum range $SR(\sum_{k=1}^{\infty} y_k$ is a closed affinite subset of Y, which may not coincide with Y (Steinitz's theorem; see [8, Theorem 2.1.1]). In case of an infinite-dimensional Banach space Y for a convergent series $\sum_n y_n$ the sum range $SR(\sum_{k=1}^{\infty} y_k)$ in general is neither closed ([9]; see also [7, Example 2.2.3 and Exercise 4.5.3]), nor affine [13].

A series $\sum_n y_n$ is called *(rearrangement) universal* in Y if $SR(\sum_{k=1}^{\infty} y_k) = Y$.

By Riemann's theorem every convergent but not unconditionally convergent series of real numbers is universal.

In [11], it is noted that by using of Riemann's theorem it is possible to find an universal series in $\mathbb{R}^d$, $d = 2, 3, \ldots$.

The existence of an universal series in $C([0, 1])$ was established in [5].

By using method of [5], in [10, pp.58-59] the existence of an universal series was proved in an arbitrary separable Banach space.

If the series $\sum_{k=1}^{\infty} \vartheta_k y_k$ converges in Y for each $(\vartheta_k) \in \{-1, 1\}^{\mathbb{N}}$ then the series $\sum_{k=1}^{\infty} y_k$ converges unconditionally; converse statement is true too provided Y is sequentially complete [12, Proposition 5.4.2].

The set $\{-1, 1\}^{\mathbb{N}}$ carries the natural product probability measure μ. We say that the series $\sum_{k=1}^{\infty} y_k$ *converges almost unconditionally in Y*, if the series $\sum_{k=1}^{\infty} \vartheta_k y_k$ converges in Y for μ-almost all $(\vartheta_k) \in \{-1, 1\}^{\mathbb{N}}$. The almost unconditionally convergent series play an important role in probability and functional analysis [12, 5.5.2].

Let $t_k = (-1)^{k+1}\frac{1}{k}$, $k = 1, 2, \ldots$. Then $\sum_k t_k$ is the universal series in $\mathbb{R}$ with properties: $\sum_{k=1}^{\infty} |t_k|^p < \infty \quad \forall p \in]1, \infty[$ and $\sum_k t_k$ converges almost unconditionally. We show that an universal series with the similar properties can be constructed in every separable Banach space.

Theorem 1.2. *Let X be a separable Banach space. Then in X there exists an universal series $\sum_i x_i$ with the following additional properties:*

$$\sum_{i=1}^{\infty} \|x_i\|^p < \infty \quad \forall p \in]1, \infty[$$

and $\sum_i x_i$ converges almost unconditionally.

Our proof of this statement uses Theorem 1.3(a) which is given next.

Let us denote by $\mathfrak{R}_{ca}$ (resp. $\mathfrak{R}_c$) the class of all real Hausdorff topological vector spaces Y which have this property: for almost every unconditionally convergent series $\sum_n y_n$ consisting of elements for Y the set $SR(\sum_{k=1}^{\infty} y_k)$ is a *closed affine subset* for Y (resp. is a *closed subset* of Y).

Theorem 1.3. *We have:*

(a) Normed spaces belong to $\mathfrak{R}_{ca}$ [1].

(b) Locally bounded Hausdorff topological vector spaces belong to $\mathfrak{R}_{ca}$ [4].

(c) Metrizable locally convex topological vector spaces belong to $\mathfrak{R}_{ca}$ [2].

It is clear that $\mathfrak{R}_{ca} \subset \mathfrak{R}_c$. We do not know whether this inclusion is strict. Also, we do not know an example of a metrizable topological vector space Y which does not belong to $\mathfrak{R}_{ca}$. Our aim is to prove the following statement.

Theorem 1.4. *Let X be a complete separable metrizable topological vector space. If $X \in \mathfrak{R}_c$, then in X there exists an universal series $\sum_i x_i$ which converges almost unconditionally.*

2. A Version of Hadwiger-Shklyarskii's Theorem

In this section, we give a convenient for our further use formulation of the result of [10, pp.58-59].

Theorem 2.1. (H. Hadwiger- O.D.Shklyarskii) *Let Y be a separable normed space, $\{y_1, y_2, \dots, y_n, \dots\}$ be a countable dense in Y set, (N_n) be a strictly increasing sequence of natural numbers such that*

$$\lim_n \frac{y_n}{N_n} = 0, \tag{2.1}$$

Write:

$$x_n = \frac{y_n}{N_n}, \; n = 1, 2, \dots,$$

$$m_0 = 0, \; m_n = \sum_{j=1}^{n} N_j, \; n = 1, 2, \dots$$

and

$$J_n = \{2m_{n-1} + 1, \dots, 2m_{n-1} + N_n, 2m_{n-1} + N_n + 1, \dots, 2m_n\}.$$

Define now a sequence $(b_i)_{i\in\mathbb{N}}$ as follows: for a fixed $i \in \mathbb{N}$ find an unique n, such that $i \in J_n$. Then $i = 2m_{n-1} + l$, where $1 \leq l \leq 2N_n$. We set, $b_i = x_n$, if l is odd and $b_i = -x_n$ otherwise.

Then the series $\sum_i b_i$ is universal.

Now from Theorem 2.1, we derive the following statement which was not noticed earlier.

Proposition 2.2. *Let Y be a separable normed space. Then in Y there exists an universal series $\sum_i b_i$ such that*

$$\sum_{i=1}^{\infty} \|b_i\|^p < \infty \quad \forall p \in]1, \infty[. \tag{2.2}$$

Proof. Let $\{y_1, y_2, \dots, y_n, \dots\}$ be a countable dense in Y set. Write:

$$K_n = \sum_{j=1}^{n} [\|y_j\|] + n, \; n = 1, 2, \dots$$

and

$$N_n = K_n^n, \; n = 1, 2, \dots$$

We have:

$$\|\frac{y_n}{N_n}\| = \frac{\|y_n\|}{N_n} \leq \frac{K_n}{N_n} = \frac{1}{K_n^{n-1}},\ n = 1, 2, \dots$$

This implies that the condition (2.1) of Theorem 2.1 is satisfied. Let $\sum_i b_i$ be the series constructed in Theorem 2.1 by means of sequences (y_n) and (N_n). According to this theorem the series $\sum_i b_i$ is universal. Let us verify that it satisfies (2.2). Fix $p \in]1, \infty[$. We have (in notation of Theorem 2.1):

$$\sum_{i=1}^{\infty} \|b_i\|^p = \sum_{n=1}^{\infty} \sum_{i \in J_n} \|b_i\|^p = 2 \sum_{n=1}^{\infty} N_n \|\frac{y_n}{N_n}\|^p = 2 \sum_{n=1}^{\infty} \frac{\|y_n\|^p}{N_n^{p-1}} \leq 2 \sum_{n=1}^{\infty} \frac{K_n^p}{N_n^{p-1}}$$

$$= 2 \sum_{n=1}^{\infty} \frac{1}{K_n^{n(p-1)-p}} = 2 \sum_{n < \frac{p+2}{p-1}} \frac{1}{K_n^{n(p-1)-p}} + 2 \sum_{n > \frac{p+2}{p-1}} \frac{1}{K_n^{n(p-1)-p}} \leq$$

$$2 \sum_{n < \frac{p+2}{p-1}} \frac{1}{K_n^{n(p-1)-p}} + 2 \sum_{n > \frac{p+2}{p-1}} \frac{1}{K_n^2} < \infty .$$

Therefore, (2.2) is satisfied too. □

Note that Proposition 2.2 is best possible in the following sense: if a series $\sum_i b_i$ in a Banach space $Y \neq \{0\}$ satisfies the condition $\sum_{i=1}^{\infty} \|b_i\| < \infty$, then $SR(\sum_i b_i)$ is singleton; hence $\sum_i b_i$ cannot be universal.

Remark 2.3. A construction of an universal series $\sum_i b_i$ in $L_2[0, 1]$ satisfy the condition $\sum_{i=1}^{\infty} \|b_i\|^2 < \infty$ was given in [3, Example 2]; a similar construction is included in [6, p. 435].

3. Proofs

We need the following lemma, the standard proof of which is ommited.

Lemma 3.1. *Let H be an infinite-dimensional separable Hilbert space, X be any complete separable metrizable topological vector space. Then there exists a continuous linear operator $T : H \to X$ such, that the image of T is dense in X, i.e., $\overline{T(H)} = X$.*

Proofs of Theorems 1.2 and 1.4. Let H be an infinite-dimensional separable Hilbert space. By Proposition 2.2 in H there exists an universal series $\sum_i b_i$ that satisfy (2.2). In case of a Hilbert space, it is well-known that the condition (2.2) (even the weaker condition $\sum_{i=1}^{\infty} \|b_i\|^2 < \infty$) implies that the series $\sum_i b_i$ converges almost unconditionally. Take now $T : H \to X$ as in Lemma 3.1 and set $x_i = Tb_i$, $i = 1, 2, \dots$ Then $\sum_i x_i$ converges almost unconditionally in X. So, by Theorem 1.3(a) (in case of Theorem 1.2) or by the assumption $X \in \mathfrak{R}_c$ (in case of Theorem 1.4), the set $SR(\sum_i x_i)$ is closed in X. Clearly,

$$SR(\sum_i x_i) \supset T(SR(\sum_i b_i)) = T(H).$$

Hence,

$$X = \overline{T(H)} \subset \overline{SR(\sum_i x_i)} = SR(\sum_i x_i).$$

Consequently, $SR(\sum_i x_i) = X$. Therefore, the second part of Theorems 1.2 and 1.4 are proved.

In case when X is a Banach space, we have:

$$\sum_{i=1}^{\infty} \|x_i\|^p \leq \|T\|^p \sum_{i=1}^{\infty} \|b_i\|^p < \infty \quad \forall p \in]1, \infty[$$

and the first part of Theorem 1.2 is proved as well. □

Acknowledgments

We are grateful to Professors G. Chelidze, S. Chobanyan, and V. Kvaratskhelia for their useful comments and criticism.

References

[1] S.A. Chobanyan, Structure of the set of sums of a conditionally convergent series in a normed spac. *Mat. Sb.* 128 (1985), No.1, 50–65 (Engl. Transl.: Mat. USSR-Sb 56(1987), 49–62)

[2] M.J. Chasco, S.A. Chobanyan, On rearrangements of series in locally convex spaces. *Michigan Math. J.* 44(1987), 607–617.

[3] V. Drobot, Rearrangements of series of functions. *Trans. Amer. Math. Soc.* 142(1969), 239–242.

[4] G. Giorgobiani, Structure of the set of sums of a conditionally convergent series in a p-normed spac. *Bull. Acad. Sci. Georgian SSR* 130(1988), 480–484.

[5] H. Hadwiger, Eine Bemerkung uber Unmordnung von Reihen reeler Funktionen. *Tohoku Math, J.* 46(1939), 22–25.

[6] B.S. Kashin and A.A. Saakian, *Orthogonal Series.* Nauka, Moscow, 1984. (Reprinted in English by AMS, 2005.)

[7] M.I. Kadets and V. M. Kadets, *Rearrangements of Series in Banach Spaces.* (in Russian) Tartu, 1988, 196 p.

[8] M.I. Kadets and V. M. Kadets, *Series in Banach Spaces.* Birkhauser Verlag, 1997.

[9] M.I. Ostrovskij, Domains of sums of conditionally convergent series in Banach spaces. (Russian) Teor. Funkts., *Funkts. Anal. Prilozh.* 46, 77-85 (1986). (English translation: Set of sums of conditionally convergent series in Banach spaces. *J. Sov. Math.* 48, No.5, 559-566 (1990)

[10] D.O. Shklyarskii, Conditionally convergent series of vectors. *Uspehi Mat. Nauk* 10(1944), 51–69.

[11] E. Steinitz, Bedingt convergente Reichen konvexe Systeme. *J. Reine Angew. Math.* 143(1913), 128–175.

[12] N. N. Vakhania, V.I. Tarieladze, S.A. Chobanyan, *Probability Distributions on Banach Spaces*. D. Reidel, Dordrecht, 1987.

[13] Wojtaszczyk, Jakub Onufry A series whose sum range is an arbitrary finite set. *Studia Math.* 171(3), 2005, 261–281.

In: Several Problems of Applied Mathematics and Mechanics ISBN: 978-1-62081-603-5
Editors: Ivane Gorgidze and Tamar Lominadze

Chapter 13

A Structural Pre-Sheaf of C^* - Algebras

M. Mumladze and Z. Zerakidze
Gori University, Gori, Georgia

Abstract

A structural pre-sheaf of C^* - algebras, which is also a co-pre-sheaf of C^*-algebras is constructed on the space $\Pr imU$ of primitive ideals of a C^* - algebra U. Also, co-chain and chain complexes with coefficients in a pre-sheaf and a co-pre-sheaf, respectively, are considered. A sheaf on $\Pr imU$ and pre-sheaves on PU and $SpaseU$ are constructed from the obtained pre-sheaf by stanstand methods.

Let, U be a C^*-algebra, FU be a set of states on this algebra [1]. Consider the universal representation $\pi_U : U \to \Im(H^U)$ of the algebra U [1], where $H^U = \underset{f \in FU}{\oplus} H^f$ and H^f is the Hilbert space associated with a state f. As we know, a universal representation is an exact one, it is an isometric embedding.

Let $U' \subset U$ be a subalgebra of the C^*-алгебры U. According to the Khan-Banach theorem, the mapping $Q : FU \to FU'$ defined by the formula $Q(f) = f|_{U'}$ is surjective.

Consider the Hilbert space $H^U = \underset{f|_{U'} \in FU'}{\oplus} H^{f|_{U'}}$. Clearly, there exists an isometric embedding $I_{U'} : H^{U'} \to H^U$ and therefore it can be assumed that $H^{U'} \subset H^U$.

Consider the projector $P_{U'} : H^U \to H^{U'}$. Let $A \in \Im(H^U)$, then $P_{U'} \circ A \circ I_{U'} \in \Im(H^{U'})$ and we have the *-морфизм $T_U : \Im(H^U) \to \Im(H^{U'})$ defined by the formula $T_U(A) = P_{U'} \circ A \circ I_{U'}$.

If $\pi_{U'}: U' \to \Im(H^{U'})$ is a universal representation, then we assume that $\tau_{U'}: U \to U'$ is the mapping defined by the formula $\tau_{U'} = \pi_{U'}^{-1} \circ T_U \circ \pi_U : U \to U'$. It is well-defined because if $A \in \pi_U(U)$, then from the construction of the universal representation it follows that $P_{U'} \circ A \circ I_{U'} \in \pi_{U'}(U')$, which means that $\tau_{U'}: U \to U'$ is well-defined.

Let $\Pr imU$ be the space of primitive ideals of the C^* -алгебры U with Jacobson topology [1]. The elements of this space are the kernels of irreducible representations of the C^*-algebra U. It is obvious that these kernels are bilateral ideals. Assume that $U' \subset U'$. Consider those irreducible representation π^U which have the form $\pi^U = \pi^{U'} \circ \tau_{U'}$, where $\pi^{U'}: U' \to \Im(H')$. The set of kernels of such representations is a subset of the space $\Pr imU$.

Let us now consider the space $\Pr imU'$ and the mapping $S: \Pr imU' \to \Pr imU$ defined by the formula $S(\beta) = \tau_{U'}^{-1}(\beta); \beta \in \Pr im, \beta = \ker \pi^U$. This mapping is an injection since if

$$\beta_1, \beta_2 \in \Pr imU'; \beta_1 \neq \beta_2, a \in \beta_1, a \notin \beta_2,$$

$$\tau_{U'}^{-1}(a) \in \tau_{U'}^{-1}(\beta_1) = \tau_{U'}^{-1}(\beta_2),$$

then $a \in \beta_2$, which is impossible. Thus our assumption is not true, $\beta_1 = \beta_2$ and we can write $\Pr imU' \subset \Pr imU$.

Assume that $O \in \Pr imU$ is an open set. It is a complement of the set F of primitive ideals, which contain some fixed subset $M \subset U$.

Thus $O = \Pr imU \setminus F$ consists of primitive ideals U of algebra, which do not contain the subset $M \subset U$.

The closure $\overline{O}$ is the set of primitive ideals, which contain some subset $M' \subset U; M' \neq M$ of the algebra U. The set $P \cap \overline{O}$ consists of primitive ideals, which contain both the set M and the set M'.

The set $\overline{O} \mid (F \cap \overline{O}) = O$ consists of primitive ideals, which contain the set M' and do not contain the set M.

Let $\alpha \in O$, then $\alpha = \ker \pi^U$, where $\pi^U: U \to \Im(H)$ is some irreducible representation. Consider the factor-algebra $U/_{U'}$, where $U' = \bigcap_{\alpha \in O} \alpha$. If $p: U \to U/_{U'}$ is a canonical projection and $\pi^{U/_{U'}}: U/_{U'} \to \Im(H)$ is some irreducible representation, then the surjectivity of p implies that the representation $\pi_{U'}^{U} = \pi^{U/U'} \circ p$ is irreducible. It is obvious that $M' \subset \ker \pi_{U'}^{U}$. Hence it follows that $\ker \pi_{U'}^{U} \in \overline{O}$.

Consider $\alpha \in O$ and the representation $\pi_\alpha \circ \overline{p} \circ p : U \to \Im(H^{U/\alpha})$, where $\overline{p} : U/U' \to U/\alpha$ is defined by the formula $\overline{p}([a]) = [[a]]$, the class $[[a]]$ is obtained by the equivalence relation $[a] \approx [b]$ if $a-b \in \alpha$, $\pi_\alpha : U/\alpha \to \Im(H^{U/\alpha})$ is the irreducible injective representation of the primitive algebra U/α.

Since $\overline{p} \circ p : U \to U/\alpha$ is a canonical projection, we have $\ker(\pi_\alpha \circ \overline{p} \circ p) = \alpha$. Therefore if $\pi^{U/U'} = \pi_\alpha \overline{p}$, then $\ker \pi_{U'}^{U} = \alpha$ and the set of primitive ideals of the algebra U/U' coincides with the set O.

Let $\pi^{U/U'} : U/U' \to \Im(H)$ be some irreducible representation. We consider the representation $\pi_{U'}^{U} = \pi^{U/U'} \circ p$ and the kernel of this representation $\ker \pi_{U'}^{U} = p^{-1}(\ker \pi^{U/U'})$.

The factor-algebra $U/p^{-1}(\ker \pi^{U/U'})$ is primitive. The *-morphism $\pi_{p^{-1}(\ker \pi^{U/U'})} : U/p^{-1}(\ker \pi^{U/U'}) \to \Im(H)$ defined by the formula

$\pi_{p^{-1}(\ker \pi^{U/U'})}([[a]]) = \pi^{U/U'}([a])$ is an irreducible injective representation. Thus $\pi_{U'}^{U} = \pi_{p^{-1}(\ker \pi^{U/U'})} \circ \overline{p} \circ p$ and hence it follows that each representation $\pi_{U'}^{U} = \pi^{U/U'} \circ p$ has the form $\pi_{U'}^{U'} = \pi_\alpha \circ \overline{p} \circ p$ and its kernel is equal to $\alpha \in O$.

From the above reasoning we have $O = \Pr im U/U'$, where $U' = \bigcap_{\alpha \in O} \alpha$.

Since U is a C^*-algebra, for each bilateral ideal $I \subset U$ the algebra U/I is a primitive module U [1]. Hence it follows that there exists a continuous *-monomorphism $\varphi : U/I \to U$, for which the equality $p \circ \varphi = id_{U/I}$ is fulfilled. Therefore it can be assumed that $U/U' \subset U$.

As we see, to each set $O \subset \Pr im U$ there corresponds a C^*-algebra $U_O \cong U/U'; U' = \bigcap_{\alpha \in O} \alpha$ such that $O = \Pr im U_O$. If $O_1 \subset O_2 \subset \Pr im U$, then $U_1' \subset U_2'$ and there exists a C^*-morphism of the algebra $U_{O_2} \cong U/U_2'; U_2' = \bigcap_{\alpha \in O_2} \alpha$ into the algebra $U_{O_1} \cong U / U_1'; U_1' = \bigcap_{\alpha \in O_1} \alpha$. We denote this C^*-morphism as follows: $\rho_{O_1}^{O_2} : U_{O_2} \to U_{O_1}$.

Clearly, if $O_1 \subset O_2 \subset O_3 \subset \mathrm{Pr}imU$, then $\rho_{O_1}^{O_2} \circ \rho_{O_1}^{O_2} = \rho_{O_2}^{O_3}$ and $\rho_O^O = Id_{U_O}$.

If, additionally, to the empty set $\Theta \subset \mathrm{Pr}imU$ we put into correspondence a trivial C^*-algebra, then on the space $\mathrm{Pr}imU$ we will define the structural pre-sheaf of C^*-algebras [2, 3]. We denote it by $\Im_U$.

The pre-sheaf $\Im_U$ has the following property: for any open subsets $O' \subset O \subset \mathrm{Pr}imU$ we have $U_O \subset U_{O'}$. This means that $\Im_U$ is a co-pre-sheaf too [2, 3]. We denote it by $\Im_U^C$.

Let us consider some open covering $\Omega = \{O_i\}$ of the space $\mathrm{Pr}imU$. To this covering there are corresponding chain and co-chain complexes with coefficients in the co-pre-sheaf $\Im_U^C$ and the pre-sheaf $\Im_U$. The modules of these chains and co-chains are respectively:

$$C_k(\Omega, \Im_U^C) = \prod_{j_0 j_1 \dots j_k} \Gamma_C(O_{j_0} \cap O_{j_1} \cap \dots \cap O_{j_k}); \partial_k : C_k(\Omega, \Im_U^C) \to C_{k-1}(\Omega, \Im_U^C);$$

$$(\partial_k f)_{j_0 j_1 \dots j_{k-1}} = \sum_{\omega} \sum_{s=0}^{k} (-1)^s r_{O_{j_o} \cap O_{j_1} \cap \dots \cap O_{j_{s-1}} \cap O_\theta \cap \dots \cap O_{j_{k-1}}}^{O_{j_0} \cap O_{j_1} \cap \dots \cap O_{j_{k-1}}} (f_{j_0 j_1 \dots j_{s-1}\ \omega\ j_s \dots j_{k-1}}),$$

$$C^k(\Omega, \Im_U) = \prod_{j_0 j_1 \dots j_k} \Gamma(O_{j_0} \cap O_{j_1} \cap \dots \cap O_{j_k}); \delta^k : C^k(\Omega, \Im_U) \to C^{k+1}(\Omega, \Im_U);$$

$$(\delta^k f)_{j_0 j_1 \dots j_{k+1}} = \sum_{s=0}^{k+1} (-1)^s \rho_{O_{j_0} \cap O_{j_1} \cap \dots \cap O_{j_{k+1}}}^{O_{j_0} \cap O_{j_1} \cap \dots \cap O_{j_{s-1}} \cap \hat{O}_{j_s} \cap O_{j_{s+1}} \cap \dots \cap O_{j_{k+1}}} (f_{j_0 j_1 \dots j_{s-1} \hat{j}_s j_{s+1} \dots j_{k+1}}).$$

where

$$O_\omega \subset \mathrm{Pr}im; O_{j_0} \cap O_{j_1} \cap \dots \cap O_{s-1} \cap O_\omega \cap O_s \cap \dots \cap O_{j_{k-1}} \neq \Theta;$$
$$r_O^W : \Im_U^C(O) \to \Im_U^C(W)\ \rho_O^W : \Im_U(W) \to \Im_U(O) \text{ when } O \subset W;\ j_0 j_1 \dots j_{j_k}$$

is ordered totality;

$$f_{j_0 j_1 \dots j_k} \in \Im_U^C(O_{j_0} \cap O_{j_1} \cap \dots \cap O_{j_k}) \text{ or } f_{j_0 j_1 \dots j_k} \in \Im_U(O_{j_0} \cap O_{j_1} \cap \dots \cap O_{j_k}).$$

Let $\{\Omega\}$ be a set of all open coverings of the space $\mathrm{Pr}imU$ and let a covering $\Omega' = \{O_l'\}$ be inscibed into $\Omega = \{O_i\}$. Fix some mapping $\theta : \{O_l'\} \to \{O_i\}$, $\theta(O_l') = O_i$ if $O_l' \subset O_i$.

Using this mapping we define the chain mapping $\theta_* : C_k(\Omega', \Im_U^C) \to C_k(\Omega, \Im_U^C)$ and the co-chain mapping $\theta^* : C^k(\Omega, \Im_U) \to C^k(\Omega', \Im_U)$ by the following formulas:

$$(\theta_* f)_{i_0 i_1 \ldots i_k} = \sum_{O'_{i_0} \cap O'_{i_1} \cap \ldots \cap O'_{i_k}} r_{O'_{l_1} \cap O'_{l_1} \cap \ldots \cap O'_{l_k}}^{\theta(O'_{l_0}) \cap \theta(O'_{l_1}) \cap \ldots \cap \theta(O'_{l_l})} (f_{l_0 l_1, \ldots l_k}).$$

$$(\theta^* f)_{l_0 l_1 \ldots l_k} = \rho_{O_{l_0} \cap O_{l_1} \cap \ldots \cap O_{l_k}}^{\theta(O_{l_0}) \cap \theta(O_{l_1}) \cap \ldots \cap \theta(O_{l_k})} (f_{i_0 i_1 \ldots i_k})$$

where $\theta(O_{l_0}) = O_{i_0}, \theta(O_{l_1}) = O_{i_1}, \ldots, \theta(O_{l_k}) = O_{i_k}$ and $O_{l_0} \cap O_{l_1} \cap \ldots \cap O_{l_k} \neq \Theta$.

It is obvious that the inverse system of chain complexes and the direct system of co-chain complexes with coefficients in a co-pre-sheaf and a pre-sheaf, respectively, are defined by the set $\{\Omega\}$ of open coverings.

Note that $C_k(\Omega, \Im_U^C); C^k(\Omega, \Im_U)$ are considered here as modules over complex numbers but not as C^*-algebras.

Assume that $\Omega = \{O^\alpha\}; O^\alpha \subset \Pr imU$ is some family of open subsets, $\{\gamma^\alpha\}$ is some consistent Ω- family of elements $\gamma^\alpha \in U_{O'}$ [4]. Such a family always exists. The set of all such consistent Ω-families makes up a C^*-алгебру. Denote it as follows: U_Ω.

Let the family $\Xi = \{G^\beta\}; G^\beta \in \Pr imU$ be inscribed into the family Ω. We denote a C^*-algebra of consistent Ξ families by U_Ξ. The morphism of C^*-algebras $\Gamma_\Xi^\Omega : U_\Omega \to U_\Xi$ is defined by the formula

$$\Gamma_\Xi^\Omega(\{\gamma^\alpha\}) = \{\rho_\Xi^\Omega(\gamma^\alpha)\}; G^\beta \subset O^\alpha, G^\beta \in \Xi, O^\alpha \in \Omega.$$

Since $\{\gamma^\alpha\}$ is a consistent Ω- family, Γ_Ξ^Ω is well-defined.

Let $\Omega_W = \{\Omega_W^\alpha\}$ be the family of all open coverings of the set $W \subset \Pr imU$, the $\{U_{\Omega_W^\alpha}; \Gamma_{\Omega_W^{\alpha'}}^{\Omega_W^\alpha} : U_{\Omega_W^\alpha} \to U_{\Omega_W^{\alpha'}}, \Omega_W^{\alpha'} \prec \Omega_W^\alpha; \Omega_W^\alpha, \Omega_W^{\alpha'}\}$ is a direct spectrum of C^* algebras.

The limit $\hat{U}_W = \varinjlim\{U_{\Omega_W^\alpha}\}$ of this direct spectrum is also a C^*-algebra [5].

Let us consider the open set $W', W' \subset W$ and the open covering Ω_W^α of the set W. The family $\Omega_{W'}^\alpha = \{O \cap W' \mid O \in \Omega_W^\alpha\}$ is the covering of the set W'. The family Ω_W^α is also the covering of the set W', therefore $\Omega_{W'}^\alpha$ is inscribed into Ω_W^α.

As shown above, there exist C^*-morphisms $\Gamma_{\Omega_{W'}^\alpha}^{\Omega_W^\alpha} : U_{\Omega_W^\alpha} \to U_{\Omega_{W'}^\alpha}$. The limit of the direct spectrum of these morphisms $\Gamma_{W'}^W : \hat{U}_W \to \hat{U}_{W'}$ is also a C^*- morphism.

It can be easily proved that the family $\{\hat{U}_W\}$ creates a sheaf on the space $\Pr imU$.

Let $SpecU$ be the spectrum of a C^*-algebra U. There exists a surjection $\phi : SpecU \to \Pr imU$ [6]. It defines a topology on the set $SpecU$. This is the weakest topology among such topologies on $SpecU$, for which ϕ is continuous.

Let now PU be the set of all pure states on U [1]. There exists a surjection $\varphi : PU \to SpecU$ [6] which also defines a topology on the set PU. This is the weakest topology among such topologies on PU, for which φ is continuous when on $SpecU$ we have the above-mentioned topology.

Topological structures on the spaces PU, $SpecU$ and $\mathrm{Pr}imU$ are not always good (in the sense of separability), but we think that the construction of such structures makes sense all the same.

The mappings $\phi, \varphi, h = \varphi \circ \phi$ are open and therefore the structures of a pre-sheaf and a co-pre-sheaf can be extended from the space $\mathrm{Pr}imU$ to PU and $SpecU$.

References

[1] A. Ya. Khelemski. Banach and poly-normed algebras. *General theory of homology representation*. Nauka, Moscow, 1989. (On Russian).

[2] R. Godeman. *Algebraic topology and the theory of sheaves*. Mir, Moscow, 1961. (On Russian).

[3] V.D. Golovin. *Homologies of analytic sheaves and the duality theory*. Nauka, Moscow, 1986. (On Russian).

[4] E. Spanier. *Algebraic topology*. Mir, Moscow, 1971. (On Russian).

[5] T.K. Kandelaki. K-Theory of C^*-algebras. *Proc. A. Razmadze Tbilisi Math. Inst. Collected Workson Topology*. 3. Metsniereba, Tbilisi, 1982. (On Georgian).

[6] J. Dixmier. C^*- *Algebras and their representations*. Nauka, Moscow, 1974. (On Russian).

In: Several Problems of Applied Mathematics and Mechanics ISBN: 978-1-62081-603-5
Editors: Ivane Gorgidze and Tamar Lominadze

Chapter 14

LEXICOGRAPHIC BIMATRIX GAME'S MIXED EXTENSION WITH CRITERIA

G. N. Beltadze
Georgian Technical University, Tbilisi

Abstract

In this paper the new concept of mixed extensional lexicographic bimatrix game $\Gamma = (\Gamma^1,...,\Gamma^r)$ of finite dimension is discussed. In this extension the players choose the components $\Gamma^1,...,\Gamma^r$ of the game Γ with given probabilistic distributions on the set of such scalar game-criteria. The concept of the situation, a situation in the coordinated strategies, an equilibrium situation with positive weights of preferences is given in a lexicographic game. The conditions of the existence of such equilibrium situation are proved in a lexicographic bimatrix game.

Keywords: lexicographic game, bimatrix game, strategy, equilibrium situation

1. Introduction

Consider a $m \times n$ bimatrix game $\Gamma = (\Gamma^1,...,\Gamma^r)$ with a matrix of payoff respectively H_1 and H_2 player's 1 and 2. Their elements are r-dimensional vectors, when they have a strong preference with the necessity of criteria. We can consider such a game as a separate class of r-dimensional of finite lexicographic noncooperative games with two players [1]:

$$\Gamma = < I, \{\chi_i\}_{i\in I}, \{H_i\}_{i\in I} > \equiv (\Gamma^1,...,\Gamma^r), \tag{1}$$

where $I = \{1,2\}$ - is the set of players, $\chi_1 = \{1,...,m\}$ and $\chi_2 = \{1,...,n\}$ are their set of strategies. The payoff functions H_1 and H_2 are written

$$H_1(i,j) = (H_1^1(i,j),...,H_1^r(i,j)) \equiv \{(a_{ij}^1,...,a_{ij}^r)\},$$
$$H_2(i,j) = (H_2^1(i,j),...,H_2^r(i,j)) \equiv \{(b_{ij}^1,...,b_{ij}^r)\}.$$

So, $H_i^1,...,H_i^r\ (i=1,2)$ are the real $m \times n$ matrices. At the same time, in situations as in pure and in mixed strategies, the vectorial payoff's meanings in R^r space is ranked $\geq^L$ or $>^L$ relations. Lexicographic strong preference $a>^Lb$ for $a=(a_1, ..., a_r)$ and $b=(b_1, ..., b_r)$ vectors means that it fulfils one of the following r conditions: 1. $a_1 > b_1$; 2. $a_1 = b_1$, $a_2 > b_2$, ...; r. $a_1 = b_1$, ..., $a_{r-1} = b_{r-1}, a_r > b_r$ and $a \geq^L b$, if $a>^Lb$ or $a = b$. This definition means, that scalar criteria $H^1,...,H^r$ are related to criteria $\Gamma^1,...,\Gamma^r$ of $\Gamma = (\Gamma^1,...,\Gamma^r)$ game, at this time $H^1,...,H^r$ are strongly ranked with decreasing preference. Game (1) is called a lexicographic bimatrix game. Therefore lexicographic bimatrix and noncooperative games are called multi-criteria games, strictly ordered with the necessity of criteria. It is known that not every lexicographic noncooperative game, including a lexicographic bimatrix game $\Gamma = (\Gamma^1,...,\Gamma^r)$, there are equilibrium situations in both pure and in mixed strategies [1]. In this paper we introduce a new concept of mixed extension when players select at random $\Gamma = (\Gamma^1,...,\Gamma^r)$ game's $\Gamma^1,...,\Gamma^r$ game-criteria. Players can negotiate beforehand about the probability distribution of $\Gamma^1,...,\Gamma^r$ game's set of criteria or define them with these criteria's own preference.

2. The Main Part

In $\Gamma = (\Gamma^1,...,\Gamma^r)$ game mixed strategies of1 and 2 players are respectively the probability distribution on the set of pure strategies $X = (x_1,...,x_m) \in \Sigma_1$ and $Y = (y_1,...,y_n) \in \Sigma_2$. The player's payoff functions in a $\bar{\Gamma} = (\bar{\Gamma}^1,...,\bar{\Gamma}^r)$ game's are written:

$$H_1(X,Y) = (\sum_{i,j=1}^{m,n} a_{ij}^1 x_i y_j,..., \sum_{i,j=1}^{m,n} a_{ij}^r x_i y_j),$$
$$H_2(X,Y) = (\sum_{i,j=1}^{m,n} b_{ij}^1 x_i y_j,..., \sum_{i,j=1}^{m,n} b_{ij}^r x_i y_j),$$

where $X = (x_1,...,x_m) \in \Sigma_1$, $Y = (y_1,...,y_n) \in \Sigma_2$ and $(X,Y) \in \Sigma_1 \times \Sigma_2$.

Let's note the set of $\bar{\Gamma} = (\bar{\Gamma}^1,...,\bar{\Gamma}^r)$ game's equilibrium situation as $\sigma(\bar{\Gamma})$. The situation $(X^*,Y^*) \in \sigma(\bar{\Gamma})$ is equilibrium if and only if

$$\mathrm{H}_1(X^*,Y^*) \geq^{\mathrm{L}} \mathrm{H}_1(i,Y^*),\ \mathrm{H}_2(X^*,Y^*) \geq^{\mathrm{L}} \mathrm{H}_2(X^*,j),\ \forall i \in \chi_1, \forall j \in \chi_2. \quad (2)$$

As we have noted above, the set $\sigma(\bar{\Gamma})$ can be empty - $\sigma(\bar{\Gamma}) = \varnothing$.

Here we introduce another version of the mixed extension of lexicographic bimatrix $\Gamma = (\Gamma^1,...,\Gamma^r)$ game. In the given game of $\Gamma^k (k = 1,...,r)$ criterium the selection of probability for the first player notes as $p_k (k = 1,...,r)$, and for the second player the selection of probability for $\Gamma^k (k = 1,...,r)$ notes as $q_k (k = 1,...,r)$. Thus $p_k \geq 0, q_k \geq 0 (k = 1,...,r)$ and $\sum p_k = 1$, $\sum q_k = 1$. Note that $P = (p_1,...,p_r)$, $Q = (q_1,...,q_r)$ and $P, Q \in \Pi$.

Also, with the importance of $\Gamma^k (k = 1,...,r)$ game we mean that $p_1 > p_2 > ... > p_r$, $q_1 > q_2 > ... > q_r$. We suppose that the selection of $\Gamma^k (k = 1,...,r)$ criteria in $\Gamma = (\Gamma^1,...,\Gamma^r)$ game takes place independently from the selection of pure and mixed strategies.

In conditions of such lexicographic $\Gamma = (\Gamma^1,...,\Gamma^r)$ game the first player's strategy is pair $S_1 = (X, P)$, and for the second player - $S_2 = (Y, Q)$.

Such kind of lexicographic bimatrix $\Gamma = (\Gamma^1,...,\Gamma^r)$ game, where the situation $S = (S_1, S_2)$, note by $\bar{\Gamma}(S)$. In game $\bar{\Gamma}(S)$ players' payoffs in the situation $S = (S_1, S_2)$ are defined as the following way:

$$\mathrm{H}_1(S_1,S_2) = \mathrm{H}_1((X,P),(Y,Q)) = (p_1 q_1 \sum_{i,j=1}^{m,n} a_{ij}^1 x_i y_j, ..., p_r q_r \sum_{i,j=1}^{m,n} a_{ij}^r x_i y_j),$$

$$\mathrm{H}_2(S_1,S_2) = \mathrm{H}_2((X,P),(Y,Q)) = (p_1 q_1 \sum_{i,j=1}^{m,n} b_{ij}^1 x_i y_j, ..., p_r q_r \sum_{i,j=1}^{m,n} b_{ij}^r x_i y_j).$$

Definition 1. The situation $(S_1^*, S_2^*) = ((X^*, P^*), (Y^*, Q^*)) \in \sigma(\bar{\Gamma}(S^*))$, if any $i \in \chi_1$ and $j \in \chi_2$ takes place thefollowing inequality:

$$\mathrm{H}_1((X^*,P^*),(Y^*,Q^*)) \geq^{\mathrm{L}} \mathrm{H}_1((i,P^*),(Y^*,Q^*)), \quad (3)$$
$$\mathrm{H}_2((X^*,P^*),(Y^*,Q^*)) \geq^{\mathrm{L}} \mathrm{H}_2((X^*,P^*),(j,Q^*)).$$

Definition 2. 2.1. The quantity $p_k q_k$ $(k = 1,...,r)$ is called the weights of preferences;

2.2. We say that the strategies S_1 and S_2 are coordinated, if any $k = 1,...,r$ or $p_k q_k > 0$ or $p_k = q_k = 0$;

2.3. We call $P=(p_1,...,p_r)$ and $Q=(q_1,...,q_r)$ coordinative strategies, and (P,Q) - the situation in coordinative strategies;

2.4. If $(S_1^*,S_2^*)\in\sigma(\bar{\Gamma}(S^*))$ and every $p_k q_k>0$ $(k=1,...,r)$, then we say that (S_1^*,S_2^*) is the equilibrium situation with positive weights of preferences, and if some $p_k=q_k=0$, then we say that (S_1^*,S_2^*) is the equilibrium situation with nonnegative weights of preferences.

Lemma. For all positive numbers $\alpha_1>0,...,\alpha_r>0$ the following inequalities are equivalent

$$(a_1,...,a_r)\,{}^{L}\!\leq (b_1,...,b_r), \tag{a}$$

$$(\alpha_1 a_1,...,\alpha_r a_r)\,{}^{L}\!\leq (\alpha_1 b_1,...,\alpha_r b_r). \tag{b}$$

Proof. The proof follows from the definition of ${}^{L}\!\leq$. Suppose we have (a), i.e. one of the following conditions takes place:

1. $a_1<b_1$; 2. $a_1=b_1$, $a_2<b_2$; ... ; r. $a_k=b_k (k=1,...,r-1)$, $a_r\leq b_r$. (a')

For positive $\alpha_1,...,\alpha_r$ follows that

2. $\alpha_1 a_1<\alpha_1 b_1$; 2. $\alpha_1 a_1=\alpha_1 b_1$, $\alpha_2 a_2<\alpha_2 b_2$;... ; r. $\alpha_k a_k=\alpha_k b_k (k=1,...,r-1)$, (b')
$\alpha_r a_r\leq \alpha_r b_r$.

From these conditions follow (b). Conversely, from (b') follows (a') and therefore fulfills (a).

Theorem. If $(S_1^*,S_2^*)=((X^*,P^*),(Y^*,Q^*))\in\sigma(\bar{\Gamma}(S^*))$ with the positive weights of preferences, then $(X^*,Y^*)\in\sigma(\bar{\Gamma})$ and vice versa, if $(X^*,Y^*)\in\sigma(\bar{\Gamma})$, then there is a situation in coordinated strategies (P^*,Q^*), that $((X^*,P^*),(Y,Q^*))\in\sigma(\bar{\Gamma}(S^*))$.

Proof. According to the condition of the theorem we have (3), i.e. for all $i\in\chi_1$ and $j\in\chi_2$ we have the inequalities

$$(p_1^* q_1^*\sum_{i,j=1}^{m,n} a_{ij}^1 x_i^* y_j^*,...,p_r^* q_r^*\sum_{i,j=1}^{m,n} a_{ij}^r x_i^* y_j^*)\,{}^{L}\!\geq (p_1^* q_1^*\sum_{j=1}^{n} a_{ij}^1 y_j^*,...,p_r^* q_r^*\sum_{j=1}^{n} a_{ij}^r y_j^*),$$

$$(p_1^* q_1^* \sum_{i,j=1}^{m,n} b_{ij}^1 x_i^* y_j, ..., p_r^* q_r^* \sum_{i,j=1}^{m,n} b_{ij}^r x_i^* y_j) \geq^L (p_1^* q_1^* \sum_{i=1}^{m} b_{ij}^1 x_i^*, ..., p_r^* q_r^* \sum_{i=1}^{m} b_{ij}^r x_i^*).$$

On the strength of positiveness $p_k q_k > 0$ $(k = 1,...,r)$ from lemma we have (2), thus $(X^*, Y^*) \in \sigma(\bar{\Gamma})$.

Now, lets suppose that $(X^*, Y^*) \in \sigma(\bar{\Gamma})$:

$$(\sum_{i,j=1}^{m,n} a_{ij}^1 x_i^* y_j^*, ..., \sum_{i,j=1}^{m,n} a_{ij}^r x_i^* y_j^*) \geq^L (\sum_{j=1}^{n} a_{ij}^1 y_j^*, ..., \sum_{j=1}^{n} a_{ij}^r y_j^*), \tag{4}$$

$$(\sum_{i,j=1}^{m,n} b_{ij}^1 x_i^* y_j^*, ..., \sum_{i,j=1}^{m,n} b_{ij}^r x_i^* y_j^*) \geq^L (\sum_{i=1}^{m} b_{ij}^1 x_i^*, ..., \sum_{i,1}^{m} b_{ij}^r x_i^*)$$

for all $i \in \chi_1$ and $j \in \chi_2$.

Let select such positive numbers $p_1^* > ... > p_r^* > 0$, $q_1^* > ... > q_r^* > 0$, that $\sum p_k^* = 1$, $\sum q_k^* = 1$. Note $P^* = (p_1^*, ..., p_r^*)$, $Q^* = (q_1^*, ..., q_r^*)$. Then on the strength of lemma from (4) we write

$$(p_1^* q_1^* \sum_{i,j=1}^{m,n} a_{ij}^1 x_i y_j^*, ..., p_r^* q_r^* \sum_{i,j=1}^{m,n} a_{ij}^r x_i y_j^*) \geq^L (p_1^* q_1^* \sum_{j=1}^{n} a_{ij}^1 y_j^*, ..., p_r^* q_r^* \sum_{j=1}^{n} a_{ij}^r y_j^*),$$

$$(p_1^* q_1^* \sum_{i,j=1}^{m,n} b_{ij}^1 x_i^* y_j, ..., p_r^* q_r^* \sum_{i,j=1}^{m,n} b_{ij}^r x_i^* y_j) \geq^L (p_1^* q_1^* \sum_{i=1}^{m} b_{ij}^1 x_i^*, ..., p_r^* q_r^* \sum_{i=1}^{m} b_{ij}^r x_i^*)$$

for all $i \in \chi_1$ and $j \in \chi_2$. This means, that $((X^*, P^*), (Y, Q^*)) \in \sigma(\bar{\Gamma}(S^*))$. The theorem is proved.

As we have noted above, vectors P^* and Q^* can be defined by coordination or with their own preferences related to $\Gamma^1, ..., \Gamma^r$. These vectors must express the weights of given criteria in certain ways. In order to find such weights we can use the method of T. Saaty about hierarchical analysis necessities of mutual comparisons of $\Gamma^1, ..., \Gamma^r$ games.

Some variants are possible. In the case of players' objective behavior, any game $\Gamma^k (k = 1,..., r-1)$ must be absolutely preferable to Γ^{k+1}, that in the quantitative estimates of T. Saaty expressed through 9.

Let's find the weight vector of the criteria $\Gamma^1, ..., \Gamma^r$ for the first player [2]:

$S_1 =$

	Γ^1	Γ^2	Γ^3	. . .	Γ^{r-1}	Γ^r
Γ^1	1	9	9^2	. . .	9^{r-2}	9^{r-1}
Γ^2	.	1	.	. . .	.	9^{r-2}
.	.	.	.	. . .	.	.
Γ^{r-1}	.	.	.	. . .	1	9
Γ^r	.	.	.	. . .	.	1

Here $\alpha^0 = (9^{r-1}, 9^{r-2}, ..., 9, 1)$, by normalizing we have P^* vector.
For example, if $\Gamma = (\Gamma^1, \Gamma^2, \Gamma^3)$, then

$S_1 =$

	Γ^1	Γ^2	Γ^3
Γ^1	1	9	81
Γ^2	.	1	9
Γ^3	.	.	1

and $\alpha^0 = (81, 9, 1)$. Since 81+9+1=91, therefore $P^* = (81/91, 9/91, 1/91)$, that is the same as Q^*.

Apart from this case, players can discuss the variants of analyses of preferences on the set $\Gamma^1, ..., \Gamma^r$. For example, one player can suppose that Γ^1 is preferable to Γ^3 very strongly (by estimating 7) or discuss other subjective demonstrative estimations.

Note. With the help of playing mechanism of ($p_1^* q_1^*, ..., p_r^* q_r^*$) vector will be chosen scalar bimatrix game from $\Gamma^1, ..., \Gamma^r$, which always exists in the equilibrium situation.

Conclusion

A new mixed distribution of a lexicographic bimatrix $\Gamma = (\Gamma^1, ..., \Gamma^r)$ game means having the first and the second player's probability distributions respectively P^* and Q^* on the set of $\Gamma^1, ..., \Gamma^r$ game-criteria. With the help P^* and Q^* the corresponding weight coefficients $p_1^* q_1^*, ..., p_r^* q_r^*$ of $\Gamma^1, ..., \Gamma^r$ games are defined. P^* and Q^* vectors can be defined by means of players' coordination or with $\Gamma^1, ..., \Gamma^r$ games' own preferences. These vectors express the given criteria's weights. In order to find them we use T. Saaty's method of hierarchical analyses. The obtained result can be generalized for n player's finite lexicographic noncooperative game.

References

[1] Beltadze G.N. Equilibrium situations in lexicographic bimatrix games. *Bulletin of the Academy of Sciences of the Georgian SSR*, 1986, 121, № 1, pp. 33-36.

[2] Beltadze G.N. The problem of multiple election in the case of multicriterional candidates. Technical University of Georgia, *Transactions*, 2009, № 4 (474), pp. 66 - 80.

In: Several Problems of Applied Mathematics and Mechanics ISBN: 978-1-62081-603-5
Editors: Ivane Gorgidze and Tamar Lominadze

Chapter 15

MATHEMATICAL MODEL OF A CLOSED ECONOMIC SYSTEM CONSIDERING DELAY FACTOR AND NEUTRAL OPTIMAL CONTROL PROBLEM

I. Ramishvili
Georgian Technical University, Tbilisi, Georgia

Abstract

Mathematical model of an economic system is presented in the form of second order neutral differential equation. Necessary conditions of optimality are established for the corresponding neutral optimal control problem.

Keywords and Phrases: Mathematical model ; delay; neutral differential equation; necessary conditions of optimality

AMS Subject Classification: 93A30; 49K21

Let $Y(t)$ be a quantity of production at $t \in J = [0, \infty)$ moment expressed in money units.

The fundamental condition of product output in the closed economic system [1] has the form

$$Y(t) = C(t) + I(t), J = [0, T], \tag{1}$$

where $C(t)$ is an apply function; $I(t)$ is an internal investment.

Let us consider the case

$$C(t) = \alpha_1(t)Y(t) \tag{2}$$

$$I(t) = \alpha_2(t)\ddot{Y}(t) + \alpha_3(t)\ddot{Y}(t-\eta) + f(t, Y(t), Y(t-\tau), \dot{Y}(t), \dot{Y}(t-\sigma), u(t)). \tag{3}$$

Here $\alpha_1(t) \in (0,1), \alpha_i(t) > 0, i = \overline{2,3}$ are continuous functions, the function $f(t, z_1, ..., z_4, u)$ is continuous on the set $J \times R_+ \times R_+ \times R \times R \times R_+$ and continuously differentiable with respect to $z_i, i = \overline{1,4}$ and u; next, η, τ, σ are positive numbers (delays), $u(\cdot) \in \Omega$ is an admissible control; Ω is the set of piecewise continuous functions $u : J \to U \subset R_+$.

From (1)-(3) formulas it follows the controlled neutral differential equation

$$\ddot{Y}(t) = \alpha(t)\ddot{Y}(t) + F(t, Y(t), Y(t-\tau), \dot{Y}(t), \dot{Y}(t-\sigma), u(t)), \tag{4}$$

where

$$\alpha(t) = \frac{\alpha_3(t)}{\alpha_2(t)}, F(t, z_1, ..., z_4, u) = \frac{1-\alpha_1(t)}{\alpha_2(t)}(t)z_1 - \frac{1}{\alpha_2(t)}f(t, z_1, ..., z_4, u).$$

On the basis of equation (4) we consider the following neutral optimal control problem with the non-fixed initial moment

$$\begin{cases} \dot{x}^1(t) = x^2(t), (x^1(t) = Y(t)) \\ \dot{x}^2(t) = \alpha(t)\dot{x}^2(t-\eta) + F(t, x^1(t), x^1(t-\tau), x^2(t), x^2(t-\sigma)), \\ t \in [t_0, t_1] \subset J, u(\cdot) \in \Omega, \end{cases} \tag{5}$$

$$x^i(t) = \varphi^i(t), t \in [-\tau_0, t_0], x^i(t_0) = x_0^i, i = 1, 2, \tag{6}$$

$$x^i(t_1) = x_1^i, i = 1, 2, \tag{7}$$

$$q(w) = \int_{t_0}^{t_1} f^0(t, x^1(t), x^2(t), u(t))dt \to \min, w = (t_0, t_1, u(\cdot)). \tag{8}$$

Here $\tau_0 = \max\{\eta, \tau, \sigma\}, \varphi^1 : [-\tau_0, T] \to R_+$ is a given continuous function; $\varphi^2 : [-\tau_0, T] \to R_+$ is a given continuously differentiable function; $x_i^1 \in R_+, x_i^2 \in R, i = 0, 1$ are fixed points; $f^0(t, z_1, z_3, u)$ is continuous on the set $J \times R_+ \times R \times R_+$ and continuously differentiable with respect to z_1, z_3.

Definition 1. Let $w = (t_0, t_1, u(\cdot)) \in W = J \times J \times \Omega, t_0 < t_1$. The function $(x^1(t), x^2(t) \in (0, \infty) \times R, t \in [-\tau_0, t_1]$ is said to be solution corresponding to the element $w \in W$, if it satisfies condition (6), while on the interval $[t_0, t_1]$ the function $x(t)$ is absolutely continuous and everywhere satisfies the equation (5).

Definition 2. The element $w \in W$ is said to be admissible if the corresponding solution $x(t)$ satisfies the condition (7).

The set of admissible elements we denote by W_0.

Definition 3. The element $\hat{w} = (\hat{t}_0, \hat{t}_1, \hat{u}(t)) \in W_0$ is said to be optimal if for an arbitrary element $w \in W$ the inequality

$$q(w_0) \leq q(w)$$

holds.

The optimal problem (5)-(8) consist in finding an optimal element.

Theorem 1. Let $\hat{w} = (\hat{t}_0, \hat{t}_1, \hat{u}(t)), \hat{t}_0 > 0$ be an optimal element and $\hat{x}^1(t), \hat{x}^2(t))$ be corresponding solution; let now $\hat{t}_0 + \tau_0 < \hat{t}_1$. Then there exists a non-trivial solution $(\psi_0(t), \psi_1(t), ..., \psi_3(t))$, where $\psi_0(t) = \psi_0 = const \leq 0, \psi_2(t) = 0, t > \hat{t}_1$, of the equation

$$\begin{cases} \dot{\psi}_1(t) = -\psi_0 \hat{f}^0_{z_1}[t] - \psi_2(t)\hat{F}_{z_1}[t] - \psi_2(t+\tau)\hat{F}_{z_2}[t+\tau], \\ \psi_2(t) = \psi_3(t) + \psi_2(t+\eta)\alpha(t+\eta), \\ \dot{\psi}_3(t) = -\psi_0 \hat{f}^0_{z_3}[t] - \psi_1(t) - \psi_2(t)\hat{F}_{z_3}[t] - \psi_2(t+\sigma)\hat{F}_{z_4}[t+\sigma], \\ t \in [\hat{t}_0, \hat{t}_1] \end{cases} \tag{9}$$

such that the following conditions are fulfilled:
1) the condition for the control function $\hat{u}(t)$

$$\psi_0 \hat{f}^0[t] + \psi_2(t)\hat{F}[t] = \max_{u \in U} \Big[\psi_0 f^0(t, \hat{x}^1(t), \hat{x}^2(t), u) + F(t, \hat{x}^1(t), \hat{x}^1(t-\tau)$$

$$, \hat{x}^2(t), \hat{x}^2(t-\sigma), u)\Big], t \in [\hat{t}_0, \hat{t}_1];$$

2) the condition for the initial moment $\hat{t}_0$

$$\psi_0 \hat{f}^0[\hat{t}_0] + \psi_1(\hat{t}_0)[x_0^2 - \dot{\varphi}^1(\hat{t}_0)] + \psi_2(\hat{t}_0-)[\alpha(\hat{t}_0)\dot{\varphi}^2(\hat{t}_0 - \eta) + \hat{F}[\hat{t}_0] - \dot{\varphi}^2(\hat{t}_0)]$$

$$+\psi_2(\hat{t}_0 + \tau-)F_1^- + \psi_2(\hat{t}_0 + \sigma-)F_2^- \leq 0;$$

3) the condition for the finally moment $\hat{t}_1$

$$\psi_0 \hat{f}^0[\hat{t}_1] + \psi_1(\hat{t}_1)x_1^2 + \psi_2(\hat{t}_1)\Big[\alpha(\hat{t}_1)\dot{x}^2(\hat{t}_1 - \eta-) + \hat{F}[\hat{t}_1]\Big] \geq 0.$$

Here

$$\hat{f}[t] = f(t, \hat{x}^1(t), \hat{x}^2(t), \hat{u}(t)), \hat{F}[t] = F(t, \hat{x}^1(t), \hat{x}^1(t-\tau), \hat{x}^2(t), \hat{x}^2(t-\sigma), \hat{u}(t));$$

$$F_1^- = F(\hat{t}_0 + \tau, \hat{x}^1(\hat{t}_0 + \tau), x_0^1, \hat{x}^2(\hat{t}_0 + \tau), \hat{x}^2(\hat{t}_0 + \tau - \sigma), \hat{u}(\hat{t}_0 + \tau-))) - F(\hat{t}_0 + \tau$$

$$, \hat{x}^1(\hat{t}_0 + \tau), \varphi^1(\hat{t}_0), \hat{x}^2(\hat{t}_0 + \tau), \hat{x}^2(\hat{t}_0 + \tau - \sigma), \hat{u}(\hat{t}_0 + \tau-))); F_2^- = F(\hat{t}_0 + \sigma, \hat{x}^1(\hat{t}_0 + \sigma)$$

$$, \hat{x}^1(\hat{t}_0 + \sigma - \tau), \hat{x}^2(\hat{t}_0 + \sigma), x_0^2, \hat{u}(\hat{t}_0 + \sigma-))) - F(\hat{t}_0 + \sigma, \hat{x}^1(\hat{t}_0 + \sigma), \hat{x}^1(\hat{t}_0 + \sigma - \tau), \hat{x}^2(\hat{t}_0 + \sigma)$$

$$, \varphi^2(\hat{t}_0), \hat{u}(\hat{t}_0 + \sigma-))).$$

Theorem 2. Let $\hat{w} = (\hat{t}_0, \hat{t}_1, \hat{u}(t)), \hat{t}_1 < T$ be an optimal element and $\hat{x}^1(t), \hat{x}^2(t))$ be corresponding solution; let now $\hat{t}_0 + \tau_0 < \hat{t}_1$. Then there exists a non-trivial solution $(\psi_0(t), \psi_1(t), ..., \psi_3(t))$, where $\psi_0(t) = \psi_0 = const \leq 0, \psi_2(t) = 0, t > \hat{t}_1$, of the equation (9) such that the condition 1) is fulfilled. Moreover,

$$\psi_0 \hat{f}^0[\hat{t}_0] + \psi_1(\hat{t}_0)[x_0^2 - \dot{\varphi}^1(\hat{t}_0)] + \psi_2(\hat{t}_0+)[\alpha(\hat{t}_0)\dot{\varphi}^2(\hat{t}_0 - \eta) + \hat{F}[\hat{t}_0] - \dot{\varphi}^2(\hat{t}_0)]$$

$$+\psi_2(\hat{t}_0 + \tau+)F_1^+ + \psi_2(\hat{t}_0 + \sigma+)F_2^+ \geq 0;$$

$$\psi_0 \hat{f}^0[\hat{t}_1] + \psi_1(\hat{t}_1)x_1^2 + \psi_2(\hat{t}_1)\Big[\alpha(\hat{t}_1)\dot{x}^2(\hat{t}_1 - \eta+) + \hat{F}[\hat{t}_1]\Big] \geq 0.$$

Here

$$F_1^+ = F(\hat{t}_0 + \tau, \hat{x}^1(\hat{t}_0 + \tau), x_0^1, \hat{x}^2(\hat{t}_0 + \tau), \hat{x}^2(\hat{t}_0 + \tau - \sigma), \hat{u}(\hat{t}_0 + \tau+))) - F(\hat{t}_0 + \tau$$

$$, \hat{x}^1(\hat{t}_0 + \tau), \varphi^1(\hat{t}_0), \hat{x}^2(\hat{t}_0 + \tau), \hat{x}^2(\hat{t}_0 + \tau - \sigma), \hat{u}(\hat{t}_0 + \tau+))); F_2^- = F(\hat{t}_0 + \sigma, \hat{x}^1(\hat{t}_0 + \sigma)$$

$$, \hat{x}^1(\hat{t}_0 + \sigma - \tau), \hat{x}^2(\hat{t}_0 + \sigma), x_0^2, \hat{u}(\hat{t}_0 + \sigma+))) - F(\hat{t}_0 + \sigma, \hat{x}^1(\hat{t}_0 + \sigma), \hat{x}^1(\hat{t}_0 + \sigma - \tau), \hat{x}^2(\hat{t}_0 + \sigma)$$

$$, \varphi^2(\hat{t}_0), \hat{u}(\hat{t}_0 + \sigma+))).$$

Theorem 3. Let $\hat{w} = (\hat{t}_0, \hat{t}_1, \hat{u}(t)), \hat{t}_i \in (0, T), i = 0, 1$ be an optimal element and $\hat{x}^1(t), \hat{x}^2(t))$ be corresponding solution; let now $\hat{t}_0 + \tau_0 < \hat{t}_1$ and the points $\hat{t}_0, \hat{t}_0 + \tau, \hat{t}_0 + \sigma$ does not belong to the set $\{\hat{t}_1 - k\eta : k = 1, 2, ...\}$. Moreover, the function $\dot{\hat{x}}^2(t - \eta)$ is continuous at the point $\hat{t}_1$ and

$$F_1^- = F_1^+ =: F_1, F_2^- = F_2^+ =: F_2.$$

Then there exists a non-trivial solution $(\psi_0(t), \psi_1(t), ..., \psi_3(t))$, where $\psi_0(t) = \psi_0 = const \leq 0, \psi_2(t) = 0, t > \hat{t}_1$, of the equation (9) such that the condition 1) is fulfilled. Moreover,

$$\psi_0 \hat{f}^0[\hat{t}_0] + \psi_1(\hat{t}_0)[x_0^2 - \dot{\varphi}^1(\hat{t}_0)] + \psi_2(\hat{t}_0)[\alpha(\hat{t}_0)\dot{\varphi}^2(\hat{t}_0 - \eta) + \hat{F}[\hat{t}_0] - \dot{\varphi}^2(\hat{t}_0)]$$

$$+\psi_2(\hat{t}_0 + \tau)F_1 + \psi_2(\hat{t}_0 + \sigma)F_2 = 0;$$

$$\psi_0 \hat{f}^0[\hat{t}_1] + \psi_1(\hat{t}_1)x_1^2 + \psi_2(\hat{t}_1)\Big[\alpha(\hat{t}_1)\dot{x}^2(\hat{t}_1 - \eta) + \hat{F}[\hat{t}_1]\Big] = 0.$$

Finally, we note that the mathematical models for various classes of economic systems taking into account delays are given [2-4], while the neutral optimal control problems with the non-fixed initial moment are considered in [4-6]

References

[1] Intriligator M. Mathematical optimization and economic theory. Englewood Cliffs, N.J. Prentico-Hall, 1971. Republished by the society for industrial and applied mathematics in the SIAMS series, *"Classics in Mathematics"*, 2002.

[2] Kharatishvili G.L.,Tadumadze T.A.Delay dynamic mathematical model of output of the production and optimal control problem, *Proc. of The 6th World Multiconference on Systemics, Cybernetics and Informatics*, Orlando, Florida , 2002 , v. VI, 10-15.

[3] Kharatishvili G.,Tadumadze T.,Gorgodze N. Dynamic mathematical model of output the production and neutral optimal control problem. Reports of enlarged session of the seminar of I. *Vekua Institute of Applied Mathematics*. v 17, 2, 2003, 36-38.

[4] Arsenashvili A., Ramishvili I., Tadumadze T. Necessary conditions of optimality for neutral variable structure optimal problems with discontinuous initial condition. *Mem.Diff. Eq. Math. Phys.* 39(2006),1-34.

[5] Gorgodze N. Necessary conditions of optimality in neutral type optimal problems with non-fixed initial moment. *Mem.Diff. Eq. Math. Phys.* 19 (2000), 150-153.

[6] Ramishvili I. The linearized maximum principle for quasi-linear neutral optimal problems with discontinuous initial condition and variable delays in controls. *Mem.Diff. Eq. Math. Phys.* 31(2004), 145-148.

In: Several Problems of Applied Mathematics and Mechanics
Editors: Ivane Gorgidze and Tamar Lominadze
ISBN: 978-1-62081-603-5

Chapter 16

APPROXIMATE SOLUTION OF UNSTEADY PROBLEM OF FREE CONVECTION WEAKLY CONDUCTING DEGREE FLUID

J. Sharikadze and Kh. Mshvenieradze
I. Vekua Institute of Applied Mathematics of Iv. Javakhishvili
Tbilisi State University, Tbilisi, Georgia

Abstract

Consideration of the approximation method of solution of unsteady boundary layer equation of free convection weakly conducting degree fluid.

Keywords: weakly conducting degree fluid, free convection, magnetic field

Equation of unsteady problem of free convection weakly conducting degree fluid are:

$$\frac{\partial u}{\partial t} + \mathrm{v}_w \frac{\partial u}{\partial y} = \frac{nk}{\rho} \left| \frac{\partial u}{\partial y} \right|^{n-1} \frac{\partial^2 u}{\partial y^2} + g\beta(T - T_\infty) - \frac{\sigma B_0^2}{\rho} u, \quad (1)$$

$$\frac{\partial T}{\partial t} + \mathrm{v}_w \frac{\partial T}{\partial y} = a \frac{\partial^2 T}{\partial y^2}, \quad (2)$$

Boundary conditions and are

$$\left.\begin{aligned} &u(y,t)\big|_{y=0} = U_w, \quad u(y,t)\big|_{y=\delta} = 0, \\ &T(y,t)\big|_{y=0} = T_w, \quad T(y,t)\big|_{y=\delta_T} = T_\infty, \\ &u(y,t)\big|_{t=0} = 0, \quad T(y,t)\big|_{t=\delta} = T_{0,} \end{aligned}\right\}, \tag{3}$$

where $u(y,t)$ - velocity of fluid, U_w - velocity of $T(y,t)$ - temperature of fluid, σ - coefficient of magnetic conducting, B_0 - tension magnetic field, ρ density of fluid, a - coefficient of electrical conductivity.

Tw_0 approximation of problem (1) – (3) are:

$$T(x,y) \approx T_1(y,t) + T_2(y,t), \tag{4}$$

where $T_1(y,t)$ is the solution to the problem:

$$\frac{\partial^2 T_1}{\partial y^2} = 0, \; T_1(y,t)\big|_{y=0} = T_w, \; T_1(y,t)\big|_{y=\delta_T} = T_\infty, \; T_1(y,t)\big|_{t=0} = T_0.$$

This solution is

$$T_1(y,t) = \frac{T_\infty - T_w}{\delta_T} y + T_w. \tag{5}$$

Solution $T_2(y,t)$ gives satisfaction to equation

$$\frac{\partial^2 T_2}{\partial y^2} = \frac{1}{a}\left(\frac{\partial T_1}{\partial t} + \mathrm{v}_w \frac{\partial T_1}{\partial y}\right), \; T_2(y,t)\big|_{y=0} = T_2(y,t)\big|_{y=\delta_T} = 0, T_2(y,t)\big|_{t=0} = 0. \tag{6}$$

Using (5) and equation (6), we get:

$$T_2(y,t) = \frac{T_w - T_\infty}{6a\delta_T^2}\frac{d\delta_T}{dt} y^3 + \frac{u_w(T_\infty - T_w)}{2a\delta_T} y^2 + \frac{(T_\infty - T_w)\left(\frac{d\delta_T}{dt} - 3\mathrm{v}_w\right)y}{6a}. \tag{7}$$

The approximation solution of problem is:

$$T(x,y) \approx T_1(y,t) + T_2(y,t) = \frac{T_w - T_\infty}{\delta_T} y + T_w + \frac{T_w - T_\infty}{6a\delta_T^2}\frac{d\delta_T}{dt} y^3$$

$$+\frac{v_w(T_\infty - T_w)}{2a\delta_T}y^2 + \frac{(T_w - T_\infty)}{6a}\left(\frac{d\delta_T}{dt} - 3v_w\right)y. \tag{8}$$

Unknown thickness of boundary δ_T defined by condition

$$\left.\frac{dT}{dy}\right|_{y=\delta_T} = 0. \tag{9}$$

From (7) equation satisfaction condition (9) is

$$2\delta_T \frac{d\delta_T}{dy} - 9v_w\delta_T - 6a = 0. \tag{10}$$

$$\left.\delta_T\right|_{t=0} = 0.$$

This equation (10) is a non-linear equation. Consider some cases, when the solution is evidently received.

1. If $v_w = \text{const}$. When from (10), we get

$$e^{\delta_T}(9v_w\delta_T + 6a)^{-\frac{2a}{3v_w}} = \frac{1}{6a^{-\frac{2a}{3v_w}}}e^{\frac{9v_w}{2}t}. \tag{11}$$

2. If $v_w = \frac{c}{9}\delta_T(t)$, where $c = \text{const}$. When

$$\delta_T(t) = \sqrt{\frac{6a}{c}}(e^{ct} - 1)^{\frac{1}{2}}. \tag{12}$$

When t is small

$$\delta_T(t) \approx \sqrt{6at}. \tag{13}$$

3. If $v_w = \frac{\alpha}{\delta_T(t)}$, where $\alpha = \text{const}$. When the solution is

$$\delta_T(t) = \sqrt{(6a + 9\alpha)t}$$

4. When velocity of suction is absent - $v_w = 0$, the solution is

$$\delta_T(t) = \sqrt{6at}\ .$$

Knowledge of thickness $\delta_T(t)$ we can determine thickness of dynamical boundary layer $\delta(t)$.

The approximation solution to dynamical problem is:

$$u(x,y) \approx u_1(y,t) + u_2(y,t) \tag{14}$$

where $u_1(y,t)$ is solution to the problem:

$$\frac{\partial^2 u_1}{\partial y^2} = 0,\ u_1\big|_{y=0} = u_w = \text{const},\ u_1\big|_{y=\delta} = 0,\ T_1(y,t)\big|_{t=0} = T_0, \tag{15}$$

a $u_2(y,t)$ is solution to the problem:

$$\frac{\partial^2 u_2}{\partial y^2} == \frac{\rho}{nk}\left|\frac{\partial u_1}{\partial y}\right|^{1-n}\left[\frac{\partial u_1}{\partial t} + v_w\frac{\partial u_1}{\partial y} - g\beta(T_1 - T_\infty) + \frac{\sigma B_0^2}{\rho}u_1\right]. \tag{16}$$

$$u_2\big|_{y=0} = 0,\ u_2\big|_{y=\delta} = 0,\ u_2\big|_{t=0} = 0. \tag{17}$$

Solution to the problem (15) is

$$u_1(y,t) = u_w\left(1 - \frac{y}{\delta(t)}\right), \tag{18}$$

and solution to the problem (16) - (17) is

$$u_2(y,t) = \frac{A_2}{6}\left(y^3 - \delta^2 y\right) - \frac{A_1}{6}\left(y^2 - \delta y\right), \tag{19}$$

where

$$A_1 = \frac{\rho(-u_w)^{1-n}}{nk}(\delta_T)^{n-1}\left[-\frac{u_w v_w}{\delta(t)} - g\beta(T_w - T_\infty) + \frac{\sigma B_0^2 u_w}{\rho}\right],$$

$$A_2 = \frac{\rho(-u_w)^{1-n}}{nk}(\delta_T)^{n-1}\left[-\frac{u_w\dot{\delta}_T}{\delta_T^2} - \frac{g\beta(T_\infty - T_w)}{\delta_T} - \frac{\sigma_0 B_0^2}{\delta_T}u_w\right].$$

Finally velocity is:

$$u(y,t) \approx u_w\left(1-\frac{y}{\delta}\right) + \frac{\rho(-u_w)^{1-n}}{6nk}(\delta)^{n-1}\left[-\frac{u_w \mathrm{v}_w}{\delta(t)} - g\beta(T_w - T_\infty) + \frac{\sigma B_0^2}{\rho}u_w\right](y^3 - \delta y y)$$

$$-\frac{\rho(-u_w)^{1-n}}{2nk}(\delta)^{n-1}\left[-\frac{u_w\dot{\delta}_T}{\delta_T^2} - \frac{g\beta(T_\infty - T_w)}{\delta_T} - \frac{\sigma_0 B_0^2 u_w}{\delta}\right](y^2 - \dot{\delta}y). \tag{20}$$

Being of thickness boundary layer $\delta(t)$ demand of condition

$$\left.\frac{\partial u}{\partial y}\right|_{y=\delta} = 0,$$

when we have

$$\frac{\rho(-u_w)^{1-n}\delta_T^2}{6nk}\delta^n\left[-\frac{\mathrm{v}_w\delta}{\delta_T^2} - \frac{g\beta(T_w - T_\infty)\delta^2}{u_w\delta_T^2} - \sigma_0 B_0\frac{\delta^2}{\delta_T^2}\right]$$

$$-\frac{\rho(-u_w)^{1-n}}{2nk}\delta^n\left[-\frac{\dot{\delta}_T\delta}{\delta_T} - \frac{g\beta(T_\infty - T_w)\delta}{u_w\delta_T} - \sigma_0 B_0^2\right] = 1. \tag{21}$$

This equation tie together two thicknesses $\delta(t)$ and $\delta_T(t)$

Let

$$\frac{\delta_T}{\delta} = \alpha, \tag{22}$$

when

$$\delta(t) = \left[\frac{\rho(-u_w)^{1-n}}{6nk}\left(\frac{3\dot{\delta}_T - \mathrm{v}_w\delta_T}{\alpha} + \frac{g\beta(T_\infty - T_w)(3\alpha - \delta_T^2)}{u_w\alpha^2} - \frac{\sigma_0 B_0^2(\delta_T^2 + \alpha^2)}{\alpha^2}\right)\right]^{\frac{1}{n}}. \tag{23}$$

Surface friction on the plate is:

$$\tau = k\left|\frac{\partial u}{\partial y}\right|_{y=0}^{n} = kG^n, \tag{24}$$

where

$$G = -\frac{u_w}{\delta_T} -$$

$$-\frac{\rho(-u_w)^{1-n}}{6nk}\delta^n\left[-\frac{u_w\left(\mathrm{v}_w\delta_T^2 + 3\dot{\delta}_T\right)}{\delta_T^2} - \frac{g\beta(T_w - T_\infty)(\delta_T\delta + 3)}{\delta_T} + \frac{\sigma_0 B_0^2 u_w\left(\delta_T^2 - 3\rho\right)}{\rho\sigma}\right]. \quad (25)$$

Coefficient of surface friction is

$$C_\tau = \frac{\tau}{\rho u_w^2} = \frac{kG^n}{\rho u_w^2}. \quad (26)$$

This expression shows that the presence of the magnetic field increases the surface friction on the plate.

References

[1] Vatazhin A. B., Lubimov G. A., Regirer S. A. *Magnetohydrodinamic flow channels.* Moscow, 1970 (in Russian).

[2] Khelmi K. A. *Flow of conducting nonnewtonian fluid in boundary layer with*

[3] Shvec M. E. The approximation solution to some problems of hydrodinamic boundary layer. *Prikl. Math. and Mech.* 1949, n. 13, 3.

[4] Dobrishman E. M. Approximation solution to some unsteady problems of boundary layer. *Prikl. Math. and Mech.* 1956, n. 20, 2.

[5] Sharikadze J. V. The flow past a plate in the presence of magnetic field. *Magnetohydrodinamica*, 1968, N 4, 53-55.

In: Several Problems of Applied Mathematics and Mechanics
Editors: Ivane Gorgidze and Tamar Lominadze
ISBN: 978-1-62081-603-5

Chapter 17

APPROXIMATE METHOD FOR SOLVING UNSTEADY ROTATION PROBLEM ON POROUS PLATE IN THE CONDUCTING FLUID WITH ACCOUNT HEAT TRANSFER IN CASE OF VARIABLE ELECTROCONDUCTIVITY

***L. Jikidze*[*] *and V. Tsutskiridze*[†]**
Georgian Technical University, Tbilisi, Georgia

Abstract

By using the method of successive approximation we have studied the unsteady problem of an infinitely rotating porous plate in a conducting fluid with account of heat transfer with variable electroconductivity $\sigma = \sigma_0 \dfrac{T}{T_\infty}$, when there is an injection of the same flow with $v_w(t)$ speed into the plate.

To determine the thickness of the dynamic and thermal boundary layers, differential equations are obtained and exact solutions were found in special cases when the injection velocity varies according to different laws and between the thicknesses of a functional dependence of the form $\delta_T(t) = \gamma\delta(t)$.

All physical characteristics are calculated.

Keywords: flow, conductivity, injection velocity, magnetic field, porosity, heat transfer

AMS Subject Classification: 76W05

[*] E-mail address: levanjikidze@yahoo.com
[†] E-mail address: b.tsutskiridze@mail.ru

1. Introduction

Injection of fluid through the plate is used to reduce the growth of unstable perturbation in the boundary layer and delaying its separation. It can also serve as an effective means of intensifying processes that use heat transfer [1, 2].

In [3] the method of successive approximation studied the unsteady problem of rotation of a porous plate in a conducting fluid with allowance for heat transfer, and in [4] and[5]we studied a similar problem, with the falling stream of the fluid on the plate with velocity components

$$v_r = ar, \quad v_\varphi = 0, \quad v_z = -2az$$

and the simultaneous rotation of the porous plate and the fluid, taking into account the magnetic field and heat transfer.

2. Main Part

In this paper we studied, by means of the method of successive approximation, an unsteady problem of the motion infinite porous plate in the conducting fluid with account heat transfer and magnetic field in a case when the electroconductivity is a function of temperature and an injection of the same flow with $v_w(t)$ speed goes into the plate.

The influence of dissipative effects on the fluid flow and heat transfer are negligibly small and the intensive injection leads to a significant reduction of the radial velocity of the fluid near the plate and the temperature difference in the main stream and the plate is relatively small.

With this in mind, to solve the problem, we use the following system of equations of unsteady motion of a conducting fluid in a uniform magnetic field and the energy equation

$$\begin{cases} \dfrac{\partial v_r}{\partial t} + v_r \dfrac{\partial v_r}{\partial r} + v_z \dfrac{\partial v_r}{\partial z} - \dfrac{v_\varphi^2}{r} = -\dfrac{1}{\rho}\dfrac{\partial p}{\partial r} + \nu\left(\Delta v_r - \dfrac{v_r}{r^2}\right) - \dfrac{\sigma B_0^2}{\rho} v_r, \\ \dfrac{\partial v_\varphi}{\partial t} + v_r \dfrac{\partial v_\varphi}{\partial r} + v_z \dfrac{\partial v_\varphi}{\partial z} + \dfrac{v_\varphi v_r}{r} = \nu\left(\Delta v_\varphi - \dfrac{v_\varphi}{r^2}\right) - \dfrac{\sigma B_0^2}{\rho} v_\varphi, \\ \dfrac{\partial v_z}{\partial t} + v_r \dfrac{\partial v_z}{\partial r} + v_z \dfrac{\partial v_z}{\partial z} = -\dfrac{1}{\rho}\dfrac{\partial p}{\partial z} + \nu \Delta v_z, \\ \dfrac{\partial v_r}{\partial r} + \dfrac{v_r}{r} + \dfrac{\partial v_z}{\partial z} = 0, \\ \lambda \dfrac{\partial^2 T}{\partial z^2} = \rho C_p \left(\dfrac{\partial T}{\partial t} + v_r \dfrac{\partial T}{\partial r} + v_z \dfrac{\partial T}{\partial z} \right), \end{cases} \tag{1}$$

where $\Delta = \frac{\partial^2}{\partial r^2} + \frac{1}{r}\frac{\partial}{\partial r} + \frac{\partial^2}{\partial z^2}$, $v_r(t), v_\varphi(t), v_z(t)$ are the components of fluid velocity, T-temperature, c_p - is heat capacity for constant pressure, λ - thermal conductivity, μ -viscosity, σ - coefficient of electroconductivity, ρ - density and B_0 - the magnetic field.

System (1) must be integrated with the following initial and boundary conditions:

$$\begin{cases} t = 0,\ v_r = v_\varphi = v_z = 0,\ T = T_w(z,0), \\ t > 0,\ z = 0,\ v_r = 0,\ v_\varphi = s\omega r,\ v_z = -v_w(t),\ T = T_w(0,t), \\ \qquad z = \infty,\ v_r = 0,\ v_\varphi = 0,\ T = T_\infty. \end{cases} \tag{2}$$

Here $v_w(t)$ -injection velocity, s-parameter rotation, $\omega(t)$ - angular velocity of the plate, T_w - temperature of the plate, and T_∞ - temperature of the fluid away from the plate. Solution of (1) is sought in the form:

$$\begin{cases} v_r = \omega_0 r f(\eta,t'),\ v_\varphi = \omega_0 r q(\eta,t'),\ v_z = \sqrt{v\omega_0}\, g(\eta,t'), \\ z = \sqrt{\frac{v}{\omega_0}}\eta,\ t' = \omega_0 t,\ \omega(t) = \omega_0\omega'(t'),\ v_w(t) = \sqrt{v\omega_0}\, v_w'(t'), \\ p = -\rho v\omega_0 p'(\eta,t'). \end{cases} \tag{3}$$

We assume that the conductivity ratio is variable depending on the temperature, in the form of:

$$\sigma = \sigma_0 \frac{T}{T_\infty} \tag{4}$$

Using (3) and (4) into (1) and for use befitting unprimed quantities, we obtain the following system of equations:

$$\begin{cases} \dfrac{\partial^2 f}{\partial \eta^2} - \dfrac{\partial f}{\partial t} - \dfrac{m^2}{T_\infty} f T = g \dfrac{\partial f}{\partial \eta} + f^2 - q^2, \\ \dfrac{\partial^2 q}{\partial \eta^2} - \dfrac{\partial q}{\partial t} - \dfrac{m^2}{T_\infty} q T = 2 f q + g \dfrac{\partial q}{\partial \eta}, \\ \dfrac{\partial p}{\partial \eta} = -\dfrac{\partial^2 g}{\partial \eta^2} + \dfrac{\partial g}{\partial t} + g \dfrac{\partial g}{\partial \eta}, \\ \dfrac{\partial g}{\partial \eta} = -2f, \\ \dfrac{\partial^2 T}{\partial \eta^2} = P_r \left(\dfrac{\partial T}{\partial t} + g \dfrac{\partial T}{\partial \eta} \right), \end{cases} \tag{5}$$

where $m^2 = \dfrac{\sigma_0 B_0^2}{\rho \omega_0}$ and $P_r = \dfrac{\mu C_p}{\lambda}$ - Prandtl's number.

To determine the thickness of the dynamic and thermal boundary layers formed in the rotating plate, with the asymptotic layers considered the layers of finite thickness, which will change over time. To determine them we use the following terms:

$$\eta = \delta(t), \ \frac{\partial q}{\partial \eta} = 0, \qquad \eta = \delta_T(t), \ \frac{\partial T}{\partial \eta} = 0. \tag{6}$$

Thus, for the solution of (5) we have the following initial and boundary conditions:

$$\begin{cases} t = 0, \ f = q = g = 0, \ T = T_w(\eta, 0), \ \delta(0) = 0, \ \delta_T = 0, \\ t > 0, \ \eta = 0, \ f = 0, \ q = s\omega(t), \ g = -v_w(t), \ T = T_w(0, t), \\ \qquad \eta = \delta(t), \ f = 0, \ q = 0, \ \dfrac{\partial q}{\partial \eta} = 0, \\ \qquad \eta = \delta_T(t), \ T = T_\infty, \ \dfrac{\partial T}{\partial \eta} = 0. \end{cases} \tag{7}$$

Problem (5)- (7) will be solved by successive approximation and let us search solutions of this problem in the form of series

$$f = \sum_{k=0}^{\infty} f_k(\eta, t), \ \ q = \sum_{k=0}^{\infty} q_k(\eta, t), \ \ g = \sum_{k=0}^{\infty} g_k(\eta, t), \ \ T = \sum_{k=0}^{\infty} T_k(\eta, t). \tag{8}$$

To determine the unknown functions required only the first two approximations. Options f_0, q_0, T_0, f_1, q_1, T_1 respectively are solutions

$$\frac{\partial^2 f_0}{\partial \eta^2}=0,\ \frac{\partial^2 q_0}{\partial \eta^2}=0,\ \frac{\partial^2 T_0}{\partial \eta^2}=0,\ \begin{cases}\eta=0, & f_0=0,\ q_0=s\omega(t),\ T_0=T_w(0,t),\\ \eta=\delta(t), & f_0=0,\ q_0=0,\\ \eta=\delta_T(t), & T_0=T_\infty,\end{cases}$$

$$\begin{cases}\dfrac{\partial^2 f_1}{\partial \eta^2}=g_0\dfrac{\partial f_0}{\partial \eta}+\dfrac{\partial f_0}{\partial t}+\dfrac{m^2}{T_\infty}f_0T_0+f_0^2-q_0^2,\\ \dfrac{\partial^2 q_1}{\partial \eta^2}=g_0\dfrac{\partial q_0}{\partial \eta}+\dfrac{\partial q_0}{\partial t}+\dfrac{m^2}{T_\infty}q_0T_0+2f_0q_0,\\ \dfrac{\partial^2 T_1}{\partial \eta^2}=P_r\left(\dfrac{\partial T_0}{\partial t}+g_0\dfrac{\partial T_0}{\partial \eta}\right),\end{cases}\quad \begin{cases}\eta=0,\ f_1=0,\ q_1=0,\ T_1=0,\\ \eta=\delta(t),\ \ f_1=0,\ q_1=0,\\ \eta=\delta_T(t),\ \ T_1=0.\end{cases}$$

and functions g_0 and g_1 determined from the expressions

$$g_0=-2\int_0^\eta f_0d\zeta-v_w(t),\qquad g_1=-2\int_0^\eta f_1d\zeta.$$

Functions f_0, q_0, g_0, T_0, f_1, q_1, g_1, T_1 are as follows:

$$f_0=0,\quad q_0=s\omega\left(1-\frac{\eta}{\delta}\right),\quad g_0=-v_w(t),\quad T_0=\frac{\theta}{\delta_T}\eta+T_w,$$

$$f_1=-\frac{s^2\omega^2}{12}\left(\frac{\eta^4}{\delta^2}-\frac{4\eta^3}{\delta}+6\eta^2-3\delta\eta\right),$$

$$q_1=s\left[\omega'\left(\frac{\eta^2}{2}-\frac{\delta}{2}\eta\right)-\left(\frac{\omega}{\delta}\right)'\left(\frac{\eta^3}{6}-\frac{\delta^2}{6}\eta\right)\right]+\frac{ms\omega}{T_\infty}\left[-\frac{\theta}{\delta_T\delta}\left(\frac{\eta^4}{12}-\frac{\delta^3}{12}\eta\right)+\right.$$
$$\left.+\left(\frac{\theta}{\delta_T}-\frac{T_w}{\delta}\right)\left(\frac{\eta^3}{6}-\frac{\delta^2}{6}\eta\right)+\frac{T_w}{2}\left(\eta^2-\delta\eta\right)\right]+\frac{s\omega v_w}{2}\left(\frac{\eta^2}{\delta}-\eta\right),$$

$$g_1=\frac{s^2\omega^2}{6}\left(\frac{\eta^5}{5\delta^2}-\frac{\eta^4}{\delta}+2\eta^3-\frac{3\delta}{2}\eta^2\right),$$

$$T_1=P_r\left[\frac{\partial}{\partial t}\left(\frac{\theta}{\delta_T}\right)\left(\frac{\eta^3}{6}-\frac{\delta_T^2}{6}\eta\right)+\left(\frac{\partial T_w}{\partial t}-v_w\frac{\theta}{\delta_T}\right)\left(\frac{\eta^2}{2}-\frac{\delta_T}{2}\eta\right)\right],$$

where $\theta=T_\infty-T_w$.

To determine the unknown thicknesses $\delta(t)$ and $\delta_T(t)$ use the condition (6) continuous transition of velocity and temperature boundary layer velocity and temperature of the external flow, assuming that they are functions of time only.

To determine the thicknesses of the dynamic and thermal boundary layers, we obtain the following system of equations:

$$\begin{cases} \left(\delta^2\right)' + \left[\left(\ln\omega\right)' + m^2 \dfrac{T_w}{T_\infty}\right]\delta^2 + \dfrac{m^2\theta}{2T_\infty}\dfrac{\delta^3}{\delta_T} + 3v_w\delta = 6, \\ \left(\delta_T^2\right)' - \left[2(\ln\theta)' + \dfrac{3}{\theta}\dfrac{\partial T_w}{\partial t}\right]\delta_T^2 + 3v_w\delta_T = \dfrac{6}{P_r}. \end{cases} \tag{9}$$

Let us consider some special cases, when it will be possible to obtain an expression $\delta_T(t)$ explicitly:

1) Let $v_w(t) = \beta_T\delta_T(t)$, where $\beta_T = const$. Then the second equation of (9), we obtain the following differential equation

$$\left(\delta_T^2\right)' - \left[2(\ln\theta)' + \frac{3}{\theta}\frac{\partial T_w}{\partial t} - 3\beta_T\right]\delta_T^2 = \frac{6}{P_r}.$$

The solution of this equation can be written as:

$$\delta_T^2(t) = \frac{6\theta^2}{P_r} e^{\int_0^t \left(\frac{3}{\theta}\frac{\partial T_w}{\partial \tau} - 3\beta_T\right)d\tau} \int_0^t \frac{1}{\theta^2(\tau)} e^{-\int_0^t \left(\frac{3}{\theta}\frac{\partial T_w}{\partial \alpha} - 3\beta_T\right)d\alpha} d\tau$$

In particular, if $\theta = const$, that

$$\delta_T(t) = \sqrt{\frac{2}{\beta_T P_r}\left(1 - e^{-3\beta_T t}\right)}.$$

If $\beta_T = \dfrac{1}{\theta(t)}\dfrac{\partial T_w}{\partial t} + \dfrac{2}{3}[\ln\theta(t)]'$, then for any $\theta(t)$ we have: $\delta_T(t) = \sqrt{\dfrac{6}{P_r}t}$.

2) If the injection velocity is chosen as

$$v_w(t) = \frac{2(\ln\theta)' + \dfrac{3}{\theta}\dfrac{\partial T_w}{\partial t}}{3}\delta_T(t) \tag{10}$$

to determine the thickness of the thermal boundary layer we obtain a simple equation

$$(\delta^2{}_T(t))' = \frac{6}{P_r} \text{ whence } \delta_T(t) = \sqrt{\frac{6}{P_r}t}\,.$$

Let us assume that $v_w(t) = \beta\delta(t)$ and thicknesses between $\delta(t)$ and $\delta_T(t)$ there is a functional relationship of the form $\delta_T(t) = \gamma\delta(t)$, where β and γ are constants. Then, to determine the dynamic boundary layer from the first equation of (9) we obtain the following differential equation

$$(\delta^2)' + \left[(\ln\omega)' + \frac{m^2}{T_\infty}\left(T_w + \frac{\theta}{2\gamma}\right) + 3\beta \right]\delta^2 = 6,$$

whose solution can be written as follows

$$\delta^2(t) = \frac{6}{\omega(t)} e^{-\int_0^t \left[3\beta + \frac{m^2}{T_\infty}\left(3\beta + \frac{\theta}{2\gamma}\right)\right] d\tau} \int_0^t \omega(\tau) e^{\int_0^t \left[3\beta + \frac{m^2}{T_\infty}\left(3\beta + \frac{\theta}{2\gamma}\right)\right] d\alpha} d\tau,$$

if $\omega = const$ and $\theta = const$, then

$$\delta(t) = \sqrt{\frac{6}{3\beta + \frac{m^2}{T_\infty}\left(T_w + \frac{\theta}{2\gamma}\right)} \left(1 - e^{-\left[3\beta + \frac{m^2}{T_\infty}\left(T_w + \frac{\theta}{2\gamma}\right)\right]t}\right)}\,.$$

If $\omega = const$ and $\beta = -\frac{m^2}{3T_\infty}\left(T_w + \frac{\theta}{2\gamma}\right)$, then $\delta(t) = \sqrt{6t}$.

If the obtained expressions $\delta(t)$ and $\delta_T(t)$ calculate the circumferential component of shear stress- $\tau_{z\varphi}$, moment of resistance to rotation of the plate -M, moment coefficient of resistance -C_M, and heat transfer coefficient- N, we have:

a) for the district component of shear stress

$$\tau_{z\varphi} = -\rho\omega s r\sqrt{\nu\omega_0}\left[\frac{2 + \delta^2(\ln\omega)' + v_w\delta}{2\delta} - \left(\frac{\omega}{\delta}\right)' \frac{\delta^2}{6\omega} + \frac{m}{T_\infty}\left(\frac{\theta\delta^2}{12T_\infty} + \frac{T_w\delta}{3}\right) \right],$$

b) for the moment of resistance of the plate

$$M = \frac{\pi\rho\omega s\sqrt{\nu\omega_0}R^4}{2}\left[\frac{2+\delta^2(\ln\omega)'+v_w\delta}{2\delta} - \left(\frac{\omega}{\delta}\right)'\frac{\delta^2}{6\omega} + \frac{m}{T_\infty}\left(\frac{\theta\delta^2}{12T_\infty} + \frac{T_w\delta}{3}\right)\right],$$

c) for the torque coefficient of resistance

$$C_M = -\frac{2\pi s}{\omega\sqrt{R_e}}\left[\frac{2+\delta^2(\ln\omega)'+v_w\delta}{2\delta} - \left(\frac{\omega}{\delta}\right)'\frac{\delta^2}{6\omega} + \frac{m}{T_\infty}\left(\frac{\theta\delta^2}{12T_\infty} + \frac{T_w\delta}{3}\right)\right],$$

d) for the heat transfer coefficient

$$N = -\frac{r}{T_w}\left\{\frac{\theta}{\delta_T} - P_r\left[\frac{\delta_T^2}{6}\frac{\partial}{\partial t}\left(\frac{\theta}{\delta_T}\right) + \frac{\delta_T}{2}\frac{\partial T_w}{\partial t} - \frac{v_w\theta}{2}\right]\right\}.$$

Conclusion

From the above formulas we can easily discern the influence of the magnetic field, velocity suction fluid, angular velocity of the plate, and Reynolds and Prandtl's numbers on the physical characteristics of the flow and heat transfer.

References

[1] Thomas A.S., Cornelius, K.K. Study slotted suction boundary layer. *Aerospace engineering*. 1983, v.1, № 1, p. 98-107.

[2] Volchkov E.P., Sinayko E.I., Terekhov V.I. *Turbulent boundary layer with suction in nonisothermal conditions*. USSR, 1979, № 2, p. 37-44.

[3] Jikidze L.A. Approximate method for solving the nonstationary problem of rotation of a porous plate in a weakly conducting liquid. *Proceedings of Tbilisi University, Mathematics, Mechanics, Astronomy*. v. 320 (30), 1995: p. 65-77.

[4] Jikidze L.A. Approximate method for solving the nonstationary problem of rotation of a porous plate in the light of the oncoming flow of weakly conducting fluid and heat transfer. *Proceedings of the International Conference "Non-classical problems of mechanics*." v. 2. Kutaisi, 2007.

In: Several Problems of Applied Mathematics and Mechanics ISBN: 978-1-62081-603-5
Editors: Ivane Gorgidze and Tamar Lominadze

Chapter 18

THERMO-ELECTROELASTICITY EQUILIBRUM TRANSTROPIC BODIES IN A SPHERICAL COORDINATE SYSTEM

D. Gorgidze, Z. Siradze, I. Ramishvili and L. Kvartskhava
Department of Engineering Mechanics of Georgian Technical University,
Tbilisi, Georgia

Abstract

By method of separation of variables, the functional solution of a number of boundary and boundary-contact problems on thermo electric elasticity equilibrium of single and multi-layer bodies, limited with coordinate surfaces of spherical system of coordinates, is constructed. The body is affected by stationary temperature and electric fields and surface perturbations. The elastic body is considered as transtropic (transversally isotropic) with surface isotropy r=const.

Keywords: elastic equilibrium; stress-strain state of a sphere; functional solution; basic equation of thermo electroelastic equilibrium; transversally isotropic elastic body; shifts and deformations

Introduction

Lamet (1) researched in detail the first study of the problem of elastic equilibrium of solid and hollow spheres in cases of specified stresses and shifts on their surfaces.

In a number of works by Lamet the results were repeated or generalized. At present there is vast literature dedicated to this problem. The stress-strain state of a sphere is investigated. In particular, papers 2 and 3 give detailed reviews of respective investigations. In these papers elastic equilibrium of homogeneous, isotropic bodies in the absence of temperature disturbance is mainly discussed. The given work presents the generalization of the problem of elastic equilibrium of a sphere in case of simplification of the method of its solution.

Simplification is achieved by construction of functional solutions for the considered class of three-dimensional boundary problems of thermal electric elasticity and by substitution of classical conditions specified on boundary surfaces with equivalent conditions.

Main Part

In a spherical system of coordinates $\boldsymbol{r}$, α, β, ($0<\boldsymbol{r}<\infty$, $0<\alpha<2\pi$, $0<\beta<\pi$), the elastic body in $\Omega = (0 < r < \infty, 0 < \alpha < 2\pi, 0 < \beta < \pi)$ domain is considered. It is supposed that the elastic body is transversally isotropic (transtropic) for which surface $r= const$ is the surface of isotropy. The body is multilayer and layers are contacting on surfaces $r= const$. The body is affected with stationary temperature and electric fields. Mass forces are absent. The basic equations of thermo electro elastic equilibrium written in the spherical system of coordinates have the following form [1,2]:

$$\text{a)}\ \frac{\partial\sigma_r}{\partial r} + \frac{1}{r\sin\beta}\frac{\partial\tau_{r\alpha}}{\partial\alpha} + \frac{1}{r}\frac{\partial\tau_{r\beta}}{\partial\beta} + \frac{2\sigma_r - \sigma_\alpha - \sigma_\beta + \tau_{r\beta}ctg\beta}{r} = 0\,,$$

$$\text{b)}\frac{\partial\tau_{\alpha r}}{\partial r} + \frac{1}{r\sin\beta}\frac{\partial\sigma_\alpha}{\partial\alpha} + \frac{1}{r}\frac{\partial\tau_{\alpha\beta}}{\partial\beta} + \frac{3\tau_{r\alpha} + 2\tau_{\alpha\beta}ctg\beta}{r} = 0\,, \qquad (1)$$

$$\text{c)}\frac{\partial\tau_{\beta r}}{\partial r} + \frac{1}{r\sin\beta}\frac{\partial\tau_{\beta\alpha}}{\partial\alpha} + \frac{1}{r}\frac{\partial\sigma_\beta}{\partial x} + \frac{(\sigma_\beta - \sigma_\alpha)\,ctg\beta + 3\tau_{r\beta}}{r} = 0.$$

$$\frac{1}{r^2}\frac{\partial D_r}{\partial r} + \frac{2}{r}D_r + \frac{1}{r\sin\beta}\frac{\partial D_\alpha}{\partial\alpha} + \frac{1}{r\sin\beta}\frac{\partial(D_\beta\sin\beta)}{\partial\beta} = 0\,. \qquad (2)$$

Equations (1) are common equations of elastic equilibrium written in the spherical system of coordinates. Equation (2) is the equation of electrostatics also written in the spherical system of coordinates.

In these equations $\sigma_r, \sigma_\alpha, \sigma_\beta$ are normal stresses, $\tau_{r\alpha} = \tau_{\alpha r}, \tau_{r\beta} = \tau_{\beta r}, \tau_{\alpha\beta} = \tau_{\beta\alpha}$ are tangential stresses, D_r, D_α, D_β are the components of the vector of electric induction $\vec{D}$ along coordinate lines r,α,β.

To these equations, generally the equations of state are added [1,2]:

$$\sigma_r = C_3(\varepsilon_{\alpha\alpha} + \varepsilon_{\beta\beta}) + C_1\varepsilon_{rr} - e_2E_r - k_{10}\boldsymbol{T}, \tau_{r\alpha} = C_4\varepsilon_{r\alpha} - e_3E_\alpha\,, \sigma_\alpha = C_2\varepsilon_{\alpha\alpha} + (C_2 - 2C_5)\varepsilon_{\beta\beta} + C_3\varepsilon_{rr} - e_1E_r - k_{20}\boldsymbol{T}, \tau_{r\beta} = C_4\varepsilon_{r\beta} - e_3E_\beta\,, \qquad (3)$$

$$\sigma_\beta = (C_2 - 2C_5)\varepsilon_{\alpha\alpha} + C_2\varepsilon_{\beta\beta} + C_3\varepsilon_{rr} - e_1E_r - k_{20}\boldsymbol{T}, \tau_{\alpha\beta} = C_5\varepsilon_{\alpha\beta}.$$

$$D_r = e_1(\varepsilon_{\alpha\alpha} + \varepsilon_{\beta\beta}) + e_2\varepsilon_{rr} + э_2E_r\,, D_\alpha = e_3\varepsilon_{r\alpha} + э_1E_\alpha\,, D_\beta = e_3\varepsilon_{r\beta} + э_1E_\beta. \qquad (4)$$

where $\varepsilon_{rr}, \varepsilon_{\alpha\alpha}, \varepsilon_\beta, \varepsilon_{r\alpha} = \varepsilon_{\alpha r}, \varepsilon_{r\beta} = \varepsilon_{\beta r}, \varepsilon_{\alpha\beta} = \varepsilon_{\beta\alpha}$ are the components of deformation tensor, E_r, E_α, E_β are the components of stress vector of electric field E, which is expressed in terms of electrostatic potential φ by formula $E=-grad\ \varphi$;$c_i=c_i(r)$,(i = 1,2,3,4,5) are elastic modules measured at direct electric field, $e_j = e_j(r)$ are piezoelectric constants, $э_1 = э_1(r)$, $э_2 = э_2(r)э = э$ are dielectric constants of permeability measured at constant deformations, $k_{10} =$

$=c_1k_1 + c_1k_2$, $k_{20} = [2(c_2 - c_5)k_2 + c_3k_1]$, $k_1 = k_1(r)$ and $k_2 = k_2(r)$ are coefficients of linear thermal expansion along r and in the direction perpendicular to r; T is temperature in electric elastic field. Shifts and deformations are connected between each other with equalities [2]:

$$\varepsilon_{rr} = \frac{\partial u}{\partial r}, \varepsilon_{\alpha\alpha} = \frac{1}{r\sin\beta}\frac{\partial w}{\partial \alpha} + \frac{u}{r} + \frac{w}{r}\,ctg\beta\,, \varepsilon_{rr} = \frac{1}{r}\frac{\partial w}{\partial \beta} + \frac{u}{r},$$
$$\varepsilon_{r\alpha} = \frac{1}{r\sin\beta}\frac{\partial u}{\partial \alpha} + \frac{\partial v}{\partial r}\frac{v}{r}, \varepsilon_{\alpha\beta} = \frac{1}{r\sin\beta}\frac{\partial w}{\partial \alpha} + \frac{1}{r}\frac{\partial v}{\partial \beta}\frac{v}{r}\,ctg\beta\,, \tag{5}$$
$$\varepsilon_{r\beta} = \frac{1}{r}\frac{\partial u}{\partial \beta} + \frac{\partial w}{\partial r}\frac{w}{r}.$$

u, v, w are components of displacement vectors $\vec{U}$ along lines tangential to coordinates $\boldsymbol{r}$, α, β.

System (1) can be rewritten in the following form:

a) $\frac{\partial}{\partial r}\left[\frac{c_3+2c_4}{c_2}r^2K + \frac{c_1c_2-(c_3+2c_4)^2}{c_2}r^2\frac{\partial u}{\partial r} - \frac{2c_4'(c_3+2c_4)}{c_2}r^2u + \left(\frac{(c_3+2c_4)k_{20}}{c_2} - k_{10}\right)r^2T + \left(e_2\frac{(c_3+2c_4)e_1}{c_2}\right)r^2\frac{\partial\varphi}{\partial r}\right] + \frac{(c_4+rc_4'+c_2c_5)}{c_2}rK + 2\frac{(c_4+rc_4'c_5)(c_3+2c_4)}{c_2}r\frac{\partial u}{\partial r} + 4\frac{(c_4+rc_4'+c_2c_5)(c_3+2c_4)}{c_2}ru - \frac{(c_4+rc_4'c_5)k_{20}}{c_2}T + 2\frac{(c_4+rc_4'c_5)e_1}{c_2}\frac{\partial\varphi}{\partial z} + e_3\Delta_2\varphi - -\frac{r}{\sin\beta}\frac{\partial B_\beta}{\partial\beta} + \frac{r}{\sin\beta}\frac{\partial(B_\alpha\sin\beta)}{\partial\alpha} = 0,$

b) $\frac{\partial K}{\partial\alpha}\sin\beta\frac{\partial B_r}{\partial\beta} + \frac{\partial(rB_\beta\sin\beta)}{\partial r} + 2(c_4c_5)\frac{1}{r}\frac{\partial u}{\partial r}2(c_4 + rc_4'c_5)\frac{v}{r}\,ctg\beta + + \frac{\partial}{\partial\alpha}\left[\left(e_3' + \frac{2e_3}{r}\right)\varphi + e_3\frac{\partial\varphi}{\partial r}\right] = 0$ (6)

c) $\frac{\partial K}{\partial\beta} - \frac{\partial(rB_\alpha)}{\partial r} + \frac{1}{\sin\beta}\frac{\partial B_r}{\partial\alpha} + 2(c_4 - c_5)\frac{1}{r}\frac{\partial u}{\partial r} - 2(c_4 + rc_4' - c_5)\frac{w}{r} +$
$$\frac{\partial}{\partial\beta}\left[\left(e_3' + +\frac{2e_3}{r}\right)\varphi + e_3\frac{\partial\varphi}{\partial r}\right] = 0$$

d) $\frac{1}{\sin\beta}\left[\frac{\partial(rB_\alpha)}{\partial\alpha} + \frac{\partial(rB_\beta\sin\beta)}{\partial\beta}\right] + c_4\frac{\partial}{\partial r}\left(\frac{1}{c_5}r^2B_r\right) = 0,$

where

$$K = c_2\left(\frac{1}{r\sin\beta}\frac{\partial v}{\partial\alpha} + 2\frac{u}{r} + \frac{1}{r}\frac{\partial w}{\partial\beta} + \frac{w}{r}\,ctg\beta\right) + (c_3 + 2c_4)\frac{\partial u}{\partial r} +$$
$$2c_4'u - k_{20}T + e_1\frac{\partial\varphi}{\partial r}, \tag{7}$$

$$B_r = c_5\frac{1}{r\sin\beta}\left(\frac{\partial w}{\partial\alpha} - \sin\beta\frac{\partial v}{\partial\beta} - v\cos\beta\right), B_\alpha = c_4\frac{1}{r}\left(\frac{\partial u}{\partial\beta} - r\frac{\partial w}{\partial r} - w\right)$$
$$B_\beta = c_4\frac{1}{r\sin\beta}\left(r\sin\beta\frac{\partial v}{\partial r} + v\sin\beta - \frac{\partial u}{\partial\alpha}\right), \tag{8}$$

$$\Delta_2 = \frac{1}{\sin\beta}\frac{\partial}{\partial\beta}\left(\sin\beta\frac{\partial}{\partial\beta}\right) + \frac{1}{\sin^2\beta}\frac{\partial^2}{\partial\alpha^2}.$$

Further we suppose that the body is weakly transtropic and weakly heterogeneous, i.e.

$$c_4 = c_5 = æ = const\,, c_1 c_2 (c_3 + 2c_4)^2 = 0\,, e_3 = const. \tag{9}$$

Considering (9), (6),(7) (8)

$$\text{a)}\ \frac{\partial}{\partial r}(e_3\Delta_2\varphi)+2rK+\frac{r}{\sin\beta}\frac{\partial B_\beta}{\partial \beta}+\frac{r}{\sin\beta}\frac{\partial(B_\alpha \sin\beta)}{\partial \alpha}=$$
$$\frac{\partial}{\partial r}\left[(k_{10}-c_2k_{20})r^2T+(ce_1-e_2)r^2\frac{\partial\varphi}{\partial r}\right]-e_3\Delta_2\varphi,$$

$$\text{b)}\ \frac{\partial K}{\partial \alpha}\sin\beta\frac{\partial B_r}{\partial \beta}+\frac{\partial(rB_\beta\sin\beta)}{\partial r}+\frac{\partial}{\partial \alpha}\left[\frac{2e_3}{r}\varphi+e_3\frac{\partial\varphi}{\partial r}\right]=0, \tag{10}$$

$$\text{c)}\ \frac{\partial K}{\partial \beta}\frac{\partial(rB_\alpha)}{\partial r}+\frac{1}{\sin\beta}\frac{\partial B_r}{\partial \alpha}+\frac{\partial}{\partial \beta}\left[\frac{2e_3}{3r}\varphi+e_3\frac{\partial\varphi}{\partial r}\right]=0,$$

$$\text{d)}\ \frac{1}{\sin\beta}\left[\frac{\partial(rB_\alpha)}{\partial \alpha}+\frac{\partial(rB_\beta\sin\beta)}{\partial \beta}\right]+c_4\frac{\partial}{\partial r}\left(\frac{1}{c_5}r^2B_r\right)=0.$$

$$K=c_2\left(\frac{1}{r\sin\beta}\frac{\partial v}{\partial \alpha}+2\frac{u}{r}+\frac{1}{r}\frac{\partial w}{\partial \beta}+\frac{w}{r}ctg\beta\right)+\sqrt{c_1c_2}\frac{\partial u}{\partial r}-k_{20}T+e_1\frac{\partial\varphi}{\partial r}, \tag{11}$$

$$B_r=æ\frac{1}{r\sin\beta}\left(\frac{\partial w}{\partial \alpha}-\sin\beta\frac{\partial v}{\partial \beta}-v\cos\beta\right),\ B_\alpha=æ\frac{1}{r}\left(\frac{\partial u}{\partial \beta}-r\frac{\partial w}{\partial r}-w\right), \tag{12}$$

$$B_\beta=æ\frac{1}{r\sin\beta}\left(r\sin\beta\frac{\partial v}{\partial r}+v\sin\beta-\frac{\partial u}{\partial \alpha}\right),$$

$$c=c(r)=\sqrt{\frac{c_1}{c_2}}\,(c_1>0, c_2>0).$$

2. Consider transtropic multilayer spherical coordinate parallelepiped $\Omega=\bigcup_{k=1}^{l}\Omega_k$ where $\Omega_k=(r_{k-1}<r<r_k, 0<\alpha<\alpha_1, \beta_0<\beta<\beta_1)$, as usual $\overline{\Omega}=\bigcup_{k=1}^{l}\overline{\Omega_k}, \overline{\Omega_k}=(r_{k-1}\le r\le r_k, 0\le\alpha\le\alpha_1, \beta_0\le\beta\le\beta_1)$. Denote surface corresponding to coordinate surface $r=r_k$ as $S_k(k=\overline{0,l})$.

Boundary and contact conditions, which appear at stating the corresponding problems, have the following form:

On surface S_0 and S_l (or the same as when $r=r_0$ or $r=r_l$):

$$\text{a)}\ \sigma_r=F_{k1}(\alpha,\beta),\ \tau_{r\alpha}=F_{k2}(\alpha,\beta),\ \tau_{r\beta}=F_{k3}(\alpha,\beta); \tag{13}$$

$$\text{b)}\ u=f_{k1}(\alpha,\beta), =f_{k2}(\alpha,\beta), w=f_{k3}(\alpha,\beta);$$
$$\text{c)}\ \sigma_r=F_{k1}(\alpha,\beta), v=f_{k2}(\alpha,\beta), w=f_{k3}(\alpha,\beta);$$

$$\text{d)}\ u=f_{k1}(\alpha,\beta), \tau_{r\alpha}=F_{k2}(\alpha,\beta), \tau_{r\beta}=F_{k3}(\alpha,\beta).$$

a) $\sigma_r = F_{k4}(\alpha,\beta)$; b) $\frac{\partial T}{\partial r} = F_{k5}(\alpha,\beta)$; c) $\frac{\partial T}{\partial r} + T = F_{k6}(\alpha,\beta)$. (14)

a) $\varphi = F_{k7}(\alpha,\beta)$; b) $\frac{\partial D_r}{\partial r} = F_{k8}(\alpha,\beta)$. (15)

In each layer when $\alpha = 0$ or $\alpha = \alpha_1$

a) $v = 0, \tau_{r\alpha} = 0, \tau_{\alpha\beta} = 0, \frac{\partial T}{\partial \alpha} = 0, D_r = 0$;

b) $\sigma_r = 0$, $w = 0$, $u = 0$, $T = 0$, $\varphi = 0$; (16)

when $\beta = \beta_0$ or $\beta = \beta_1$

a) $w = 0, \tau_{r\beta} = 0, B_r = 0, \frac{\partial T}{\partial \beta} = 0, D_\beta = 0$

b) $K = 0$, $v = 0$, $u = 0$, $T = 0$, $\varphi = 0$ (17)

On contact surfaces S_k (k=1,2,…,l-1; S_k is the contact surface of layer number k contacting with layer k+1) are given conditions:

$$T^k - T^{k+1} = F_1(\alpha,\beta)\,, \frac{\partial T^k}{\partial r} - \frac{\partial T^{k+1}}{\partial r} = \tilde{F}_1(\alpha,\beta) \quad (18)$$

$$\varphi^k - \varphi^{k+1} = F_2(\alpha,\beta)\,, D_r^k - D_r^{k+1} = \tilde{F}_2 F_2(\alpha,\beta) \quad (19)$$

a) $u^k - u^{k+1} = f_1(\alpha,\beta)\,, \sigma_r^k - \sigma_r^{k+1} = F_3(\alpha,\beta), v^k - v^{k+1} = f_2(\alpha,\beta)$

$$w^k - w^{k+1} = f_3(\alpha,\beta)\,, \tau_{r\alpha}^k - \tau_{r\alpha}^{k+1} = F_4(\alpha,\beta), \tau_{r\beta}^k - \tau_{r\beta}^{k+1} = F_5(\alpha,\beta)$$

b) $u^k - u^{k+1} = f_1(\alpha,\beta)\,, \sigma_r^k - \sigma_r^{k+1} = F_3(\alpha,\beta), \tau_{r\alpha}^k = F_4(\alpha,\beta)$,

$$\tau_{r\alpha}^{k+1} = F_5(\alpha,\beta), \tau_{r\beta}^k = F_5(\alpha,\beta), \tau_{r\beta}^{k+1} = F_6(\alpha,\beta). \quad (20)$$

c) $\sigma_r^k = F_3(\alpha,\beta)\sigma_r^{k+1} = F_4(\alpha,\beta), v^k - v^{k+1} = f_2(\alpha,\beta)$

$$w^k - w^{k+1} = f_3(\alpha,\beta)\,, \tau_{r\alpha}^k - \tau_{r\alpha}^{k+1} = F_5(\alpha,\beta), \tau_{r\beta}^k - \tau_{r\beta}^{k+1} = F_6(\alpha,\beta)$$

The conditions imposed on the stated functions shall be referred to later; here we just note that these functions are such that on the edges of spherical coordinate parallelepiped the conditions of matching are satisfied. Technical interpretation of boundary conditions (16) and (17) is given in [3].

3. In this paragraph temperature problem in spherical system of coordinates is set forth. The statement of this problem is based on the results of [3]. In the first paragraph T indicates temperature in elastic body satisfying the equation:

$$\Delta_2 T + \frac{1}{\lambda_2}\frac{1}{r^2}\frac{\partial}{\partial r}\left(\lambda_1 r^2 \frac{\partial T}{\partial r}\right) = 0 \tag{21}$$

and boundary conditions (14), (16) and (17). $\lambda_1=\lambda_1(r)$ and $\lambda_2=\lambda_2(r)$ are coefficients of heat conductance along $\boldsymbol{r}$ and β in the direction perpendicular to r.

If the medium is homogeneous, i.e. when $\lambda_1=$ const and $\lambda_2=$ const, equation (21) takes the form:

$$\Delta_2 T + \lambda_0 \frac{1}{r^2}\frac{\partial}{\partial r}\left(r^2 \frac{\partial T}{\partial r}\right) = 0 \tag{22}$$

where $\lambda_0=\lambda_1/\lambda_2$. Using the Fourier method (method of variables separation), function T in the area $\Omega=\{\boldsymbol{r}_0<\boldsymbol{r}<\boldsymbol{r}_1,\ 0<\alpha<\alpha_1,\ \beta_0<\beta<\beta_1\}$ can be presented in the form:

$$T= t_0 + t_1\frac{1}{r} + \sum_{n=1}^{\infty}\sum_{m=0}^{\infty}\left[\mathrm{A}_{\mathrm{Tmn}}\left(\frac{r_0}{r}\right)^{n_t+1} + \mathrm{B}_{\mathrm{Tmn}}\left(\frac{r}{r_1}\right)^{n_t+1}\right]\Psi_{\mathrm{mn}}(\alpha,\beta) \tag{23}$$

Here t_0, t_1, A_{Tmn}, B_{Tmn}, n_t are constants. $2\sqrt{\lambda_0}n_t = \sqrt{\lambda_0 + 4\tilde{n}(\tilde{n}+1)}$, $\tilde{n} = \tilde{n}(m,n)$ (when $\lambda_0 = 1$ and $n_t = \tilde{n}$). $\Psi_{\mathrm{mn}}(\alpha,\beta)$is nontrivial solution of Sturm-Liouville problem [5].

$$r^2\Delta_2\Psi_{\mathrm{mn}} + \tilde{n}(\tilde{n}+1)\Psi_{\mathrm{mn}} = 0, \tag{24}$$

when= α_i: a) $\Psi_{\mathrm{mn}} = 0$ or b) $\dfrac{\partial\Psi_{\mathrm{mn}}}{\partial\alpha} = 0,$ (25)

when= β_i: a) $\Psi_{\mathrm{mn}} = 0$ or b) $\dfrac{\partial\Psi_{\mathrm{mn}}}{\partial\beta} = 0.$ (26)

Function $\Psi_{mn}(\alpha,\beta)$ can be presented in the following form:

$$\Psi_{\mathrm{mn}}(\alpha,\beta) = \mathrm{A}_{\alpha}(m)\cdot \mathrm{B}_{\beta}(m,n).$$

Expressions for function $\mathrm{A}_{\alpha}(m), \mathrm{B}_{\beta}(m,n)$ may have the following form:

If conditions (16a) are satisfied:

$$\mathrm{A}_{\alpha}(m)=\cos\left(\frac{\pi m}{\alpha_1}\cdot\alpha\right);$$

when (16b)

$$\mathrm{A}_{\alpha}(m)=\sin\left(\frac{\pi m}{\alpha_1}\cdot\alpha\right).$$

If when $\alpha = 0$ (16b) and when $\alpha = \alpha_1$ (16a):

$$\mathrm{A}_{\alpha}(m)=\sin\left(\frac{\pi(2m+1)}{2\alpha_1}\cdot\alpha\right);$$

while if $\alpha = 2\pi$

$$\mathrm{A}_{\alpha}(m)=a_{1m}\cos(m\cdot\alpha)+a_{2m}\sin(m\cdot\alpha)$$

where $a_{1m}, a_{2m}$are constants.

2) When $0< \beta_0 < \beta_1 < \pi$ and the conditions (17a) are satisfied:

$$\mathrm{B}_{\beta}(m,n) = P_{\tilde{n}}^{-\tilde{m}}(\cos\beta)\frac{d}{d\beta}Q_{\tilde{n}}^{-\tilde{m}}(\cos\beta_0) - Q_{\tilde{n}}^{-\tilde{m}}(\cos\beta)\frac{d}{d\beta}P_{\tilde{n}}^{-\tilde{m}}(\cos\beta_0)$$

where $P_{\tilde{n}}^{-\tilde{m}}(\cos\beta), Q_{\tilde{n}}^{-\tilde{m}}(\cos\beta)$ are the first and second kindLegandre functions, $\tilde{m}=\frac{\pi m}{\alpha_1}$ or $\tilde{m}=\frac{\pi(2m+1)}{2\alpha_1}$ or $\tilde{m}=m$ and $\tilde{n}$is the root ofn equation number:

$$\frac{d\,P_{\tilde{n}}^{-\tilde{m}}(\cos\beta)}{d\beta}\frac{dQ_{\tilde{n}}^{-\tilde{m}}(\cos\beta_0)}{d\beta} - \frac{dQ_{\tilde{n}}^{-\tilde{m}}(\cos\beta)}{d\beta}\frac{dP_{\tilde{n}}^{-\tilde{m}}(\cos\beta_0)}{d\beta}=0.$$

while if conditions (17b) are satisfied, then $\mathrm{B}_{\beta}(m,n) = P_{\tilde{n}}^{-\tilde{m}}(\cos\beta)Q_{\tilde{n}}^{-\tilde{m}}(\cos\beta_0) - -Q_{\tilde{n}}^{-\tilde{m}}(\cos\beta)P_{\tilde{n}}^{-\tilde{m}}(\cos\beta_0)$ and $\tilde{n}$is the root of equationnumber n:

$$P_{\tilde{n}}^{-\tilde{m}}(\cos\beta_1)Q_{\tilde{n}}^{-\tilde{m}}(\cos\beta_0) - -Q_{\tilde{n}}^{-\tilde{m}}(\cos\beta_1)P_{\tilde{n}}^{-\tilde{m}}(\cos\beta_0)=0$$

while if $\beta = \beta_0$, the conditions (17b) are satisfied and if $\beta = \beta_1$, the conditions (17a) are satisfied, then:

$$\mathrm{B}_{\beta}(m,n) = P_{\tilde{n}}^{-\tilde{m}}(\cos\beta)Q_{\tilde{n}}^{-\tilde{m}}(\cos\beta_0) - -Q_{\tilde{n}}^{-\tilde{m}}(\cos\beta)P_{\tilde{n}}^{-\tilde{m}}(\cos\beta_0)$$

and $\tilde{n}$ is the root ofequation number n.

$$Q_{\tilde{n}}^{-\tilde{m}}(\cos\beta_0)\frac{d}{d\beta}P_{\tilde{n}}^{-\tilde{m}}(\cos\beta_1) - P_{\tilde{n}}^{-\tilde{m}}(\cos\beta_0)\frac{d}{d\beta}Q_{\tilde{n}}^{-\tilde{m}}(\cos\beta_1)=0$$

and when $\beta = \beta_0$ the conditions (17a) are fulfilled, while when $\beta = \beta_1$ (17b) then

$$\mathrm{B}_{\beta}(m,n) = P_{\tilde{n}}^{-\tilde{m}}(\cos\beta)Q_{\tilde{n}}^{-\tilde{m}}(\cos\beta_1) - -Q_{\tilde{n}}^{-\tilde{m}}(\cos\beta)P_{\tilde{n}}^{-\tilde{m}}(\cos\beta_1)$$

and when $\tilde{n}$ is the root of equation number n:

$$Q_{\tilde{n}}^{-\tilde{m}}(\cos\beta_1)\frac{d}{d\beta}P_{\tilde{n}}^{-\tilde{m}}(\cos\beta_0) - P_{\tilde{n}}^{-\tilde{m}}(\cos\beta_1)\frac{d}{d\beta}Q_{\tilde{n}}^{-\tilde{m}}(\cos\beta_0)=0.$$

If $\beta = \beta_0$, $\beta \neq \frac{\pi}{2}$ and $\beta \neq \pi$, then from the limitation condition, in all cases (and there are only two)

$$\mathrm{B}_\beta(m,n) = P_{\tilde{\mathrm{n}}}^{-\tilde{\mathrm{m}}}(\cos\beta)$$

where instead of upper index (- $\tilde{\mathrm{m}}$) in case $\tilde{m}$ of whole $\tilde{\mathrm{m}}$ is taken.

In all considered cases when $\tilde{m} = p - \frac{1}{2}$,p =1,2,3,..., where $\mathrm{B}_\beta(m,n)$ is elementary function [4].

3) If $\beta_0 = 0$ and $\beta_1 = \frac{\pi}{2}$, then

$$\mathrm{B}_\beta(m,n) = P_{\tilde{m}+p}^{-\tilde{m}}(\cos\beta)$$

where or p=0,2,4....- p=1,3,5,....
and if $\beta_0 = 0$and $\beta_1 = \pi$, then

$$\mathrm{B}_\beta(m,n) = P_{\tilde{m}+p}^{-\tilde{m}}(\cos\beta)$$

where =0,1.2,3,4... . In both cases$P_{\tilde{m}+p}^{-\tilde{m}}(\cos\beta) = (-1)^n P_{\tilde{m}+p}^{-\tilde{m}}[\cos(\pi-\beta)]$ is an elementary function. And, in the end, if m is an integer number, the sign of upper index is changed $\mathrm{B}_\beta(m,n)$, and the function itself is polynomial p of the degree relative to $\cos\beta$[4].

If the medium is thermally inhomogeneous $\left(\lambda_1 = \lambda_1(r)\text{and } \lambda_2 = \lambda_2(r)\right)$, then

$$T = \eta_{T_0}(r) + \sum_{n=1}^{\infty}\sum_{m=0}^{\infty}\eta_{Tmn}(r)\Psi_{mn}(\alpha,\beta)$$

and $\eta_{T_0}(r)$, $\eta_{Tmn}(r)$ are the solution of following equations:

$$\frac{d}{dr}\left(\lambda_1 r^2 \frac{d\eta_{T_0}(r)}{dr}\right)=0,$$

$$\frac{1}{r^2}\frac{d}{dr}\left(\lambda_1 r^2 \frac{d\eta_{Tmn}(r)}{dr}\right) - \tilde{n}(\tilde{n}-1)\eta_{Tmn}(r) = 0.$$

4. For convenience of construction of solutions of boundary and boundary-contact problems, the conditions (13) and (20) are substituted with the following conditions:

$$\text{a) } \sigma_r = F_{k1}(\alpha,\beta),\ \Gamma_1(\tau_{r\alpha},\tau_{r\beta}) = \tilde{F}_{k2}(\alpha,\beta),\ \Gamma_2(\tau_{r\alpha},\tau_{r\beta}) = \tilde{F}_{k3}(\alpha,\beta)$$

$$\text{b) } u = f_{k1}(\alpha,\beta),\ \Gamma_1(v,w) = \tilde{f}_{k2}(\alpha,\beta),\ \Gamma_2(v,w) = \tilde{f}_{k3}(\alpha,\beta)$$

$$\text{c) } \sigma_r = F_{k1}(\alpha,\beta),\ \Gamma_1(v,w) = \tilde{f}_{k2}(\alpha,\beta),\ \Gamma_2(v,w) = \tilde{f}_{k3}(\alpha,\beta) \quad (27)$$

$$\text{d) } u = f_{k1}(\alpha,\beta),\ \Gamma_1(\tau_{r\alpha},\tau_{r\beta}) = \tilde{F}_{k2}(\alpha,\beta),\ \Gamma_2(\tau_{r\alpha},\tau_{r\beta}) = \tilde{F}_{k3}(\alpha,\beta)$$

$$\text{a) } u^k - u^{k+1} = f_1(\alpha,\beta),\ \sigma_r^k - \sigma_r^{k+1} = F_3(\alpha,\beta),\ \Gamma_1(v^k,w^k) - \Gamma_1(v^{k+1},w^{k+1}) = \tilde{f}_2(\alpha,\beta),$$

$$\Gamma_2(v^k,w^k) - \Gamma_2(v^{k+1},w^{k+1}) = \tilde{f}_3(\alpha,\beta),\ \Gamma_1(\tau_{r\alpha}^k,\tau_{r\beta}^k) - \Gamma_1(\tau_{r\alpha}^{k+1},\tau_{r\beta}^{k+1}) = \tilde{F}_4(\alpha,\beta),$$

$$\Gamma_2(\tau_{r\alpha}^k,\tau_{r\beta}^k) - \Gamma_2(\tau_{r\alpha}^{k+1},\tau_{r\beta}^{k+1}) = \tilde{F}_5(\alpha,\beta)$$

$$\text{b) } u^k - u^{k+1} = f_1(\alpha,\beta),\ \sigma_r^k - \sigma_r^{k+1} = F_3(\alpha,\beta),\ \Gamma_1(\tau_{r\alpha}^k,\tau_{r\beta}^k) = \tilde{F}_4(\alpha,\beta),$$

$$\Gamma_2(\tau_{r\alpha}^k,\tau_{r\beta}^k) = \tilde{F}_5(\alpha,\beta),\ \Gamma_1(\tau_{r\alpha}^{k+1},\tau_{r\beta}^{k+1}) = \tilde{F}_6(\alpha,\beta),$$
$$\Gamma_2(\tau_{r\alpha}^{k+1},\tau_{r\beta}^{k+1}) = \tilde{F}_7(\alpha,\beta). \tag{27'}$$

$$\text{c) } \sigma_r^k = F_3(\alpha,\beta)\sigma_r^{k+1} = F_4(\alpha,\beta),\ \Gamma_1(v^k,w^k) - \Gamma_1(v^{k+1},w^{k+1}) = \tilde{f}_2(\alpha,\beta),$$

$$\Gamma_2(v^k,w^k) - \Gamma_2(v^{k+1},w^{k+1}) = \tilde{f}_3(\alpha,\beta),\ \Gamma_1(\tau_{r\alpha}^k,\tau_{r\beta}^k) - \Gamma_1(\tau_{r\alpha}^{k+1},\tau_{r\beta}^{k+1}) = \tilde{F}_4(\alpha,\beta),$$
$$\Gamma_2(\tau_{r\alpha}^k,\tau_{r\beta}^k) - \Gamma_2(\tau_{r\alpha}^{k+1},\tau_{r\beta}^{k+1}) = \tilde{F}_5(\alpha,\beta)$$

where

$$\Gamma_1(\eta_1,\eta_2) = \frac{1}{\sin\beta}\left[\frac{\partial\,\eta_1}{\partial a} + \frac{\partial(\eta_2 \sin\beta)}{\partial\beta}\right],$$

$$\Gamma_2(\eta_1,\eta_2) = \frac{1}{\sin\beta}\left[\frac{\partial\,\eta_2}{\partial a} + \frac{\partial(\eta_1 \sin\beta)}{\partial\beta}\right],$$

where $\eta_1 = \tau_{r\alpha}$ or $\eta_1 = v, \eta_2 = \tau_{r\beta}$ or $\eta_2 = w$. It is supposed that functions $F_1(\alpha,\beta)$, $F_2(\alpha,\beta)$, $F_3(\alpha,\beta)$, $F_4(\alpha,\beta)$, $F_5(\alpha,\beta)$, $F_6(\alpha,\beta)$, $F_7(\alpha,\beta)$, $F_{k2}(\alpha,\beta)$, $F_{k3}(\alpha,\beta)$ themselves, functions $F_{k1}(\alpha,\beta)$, $f_{k2}(\alpha,\beta)$, $f_{k3}(\alpha,\beta)$, $F_3(\alpha,\beta)$, $f_2(\alpha,\beta)$, $f_3(\alpha,\beta)$, $F_1(\alpha,\beta)$, $F_3(\alpha,\beta)$, together with their first and second derivatives and functions $f_{k1}(\alpha,\beta)$, $f_1(\alpha,\beta)$, together with their first and second derivatives, disintegrate into uniformly convergent Fourier series by their Eigen functions of problems (24), (25) and (26). About equivalence of boundary conditions (13) and (27) we can say the following: equivalence of conditions and (27) are studied analogously as was given in work [3].

5. From (10b,c,d) we have:

$$\Delta\,(rB_r) = 0, \tag{28}$$

where $\Delta = \Delta_r + \frac{1}{r^2}\frac{\partial}{\partial r}\left[r^2\frac{\partial()}{\partial r}\right] = \Delta_2 + \frac{\partial^2}{\partial r^2} + \frac{2}{r}\frac{\partial}{\partial r}$. From boundary conditions (16) and (17) it follows that on lateral boundaries $\alpha=0$, $\alpha = \alpha_1$ and $\beta = \beta_0$, $\beta = \beta_1$ of coordinative parallelepiped or the function B_r or its normal derivative are equal to zero, while on surfaces $r = r_0$ or $r = r_l$, from (27) we get

$$\Gamma_2(\tau_{r\alpha},\tau_{r\beta}) = r\frac{\partial B_r}{\partial r} \text{ or } \Gamma_2(v,w) = \frac{r}{æ}B_r.$$

Thus, for function B_r we get the following problem of mathematical physics: it is necessary to determine function from equation (28) when on the surface of domain of its determination, the function itself or its normal derivative or on the part of the surface the function itself and on the other part its normal derivative are given.

Using Fourier method, represent function B_r in the following form:

$$B_r = a_{10}\frac{1}{r} + a_{11}\frac{1}{r^2} + \sum_{m=1}^{\infty}\sum_{n=1}^{\infty}\left[A_{Bmn}\left(\frac{r_0}{r}\right)^{\tilde{n}_1+1} + B_{Bmn}\left(\frac{r}{r_1}\right)^{n_1}\right]\overline{\Psi}_{mn}(a,\beta), \quad (29)$$

where a_{10} ,a_{11} are constants, $\overline{\Psi}_{mn}(a,\beta)$ is the solution of the following Sturm-Liouville problem [5].

$$r^2\Delta_2\overline{\Psi}_{mn}(a,\beta) + \tilde{n}(\tilde{n} + 1)\overline{\Psi}_{mn}(a,\beta) = 0;$$

$$\alpha{=}\alpha_i\colon a)\overline{\Psi}_{mn}(a,\beta) = 0 \ or \ b)\frac{\partial\overline{\Psi}_{mn}(a,\beta)}{\partial\alpha} = 0;$$

$$\beta{=}\beta_i\colon a)\overline{\Psi}_{mn}(a,\beta) = 0 \ or \ b)\frac{\partial\overline{\Psi}_{mn}(a,\beta)}{\partial\beta} = 0.$$

If when α =0 and $\alpha = \alpha_1$ the conditions (16b) are given, while when $\beta = \beta_0$ and $\beta = \beta_1$, the conditions (17b) are constants and a_{10} and a_{11}cannot differ from zero (in all other cases a_{10} and a_{11} are equal to zero).

Indeed, if

$$\int_0^{\alpha_1}\int_{\beta_0}^{\beta_1} B_r r^2 \sin\beta \, d\alpha \, d\beta = 0$$

(this condition is easily verified) it follows that a_{10} =0 and a_{11} =0.

Finally, for the considered boundary-value problems

$$B_r = \sum_{m=1}^{\infty}\sum_{n=1}^{\infty}\left[A_{Bmn}\left(\frac{r_0}{r}\right)^{\tilde{n}_1+1} + B_{Bmn}\left(\frac{r}{r_1}\right)^{n_1}\right]\overline{\Psi}_{mn}(a,\beta). \quad (30)$$

From (25) it follows that function, without limiting the generalities, can be presented in the following form:

$$B_r = \frac{\partial^2(r\Psi_0)}{\partial r^2} \quad (31)$$

Ψ_0 is the solution of equation $\Delta\Psi_0 = 0$and it can be presented:

$$\Psi_0 = \sum_{m=1}^{\infty}\sum_{n=1}^{\infty}\left[A_{Bmn}\left(\frac{r_0}{r}\right)^{\tilde{n}_1+1} + B_{Bmn}\left(\frac{r}{r_1}\right)^{n_1}\right]\overline{\Psi}_{mn}(a,\beta). \quad (32)$$

Theorem. If function $\Psi_0(r,\alpha,\beta), \Psi_1(r,\alpha,\beta), \Psi_2(r,\alpha,\beta)$ and φ represent the solution of the corresponding equation system:

$$a)\ \Delta\Psi_0 = 0,$$

$$b)\ \frac{\partial}{\partial r}\left(cr^2\frac{\partial(r\Psi_1)}{\partial r}\right) - 2r\frac{\partial(r\Psi_1)}{\partial r} + r^3\Delta_2\Psi_1 + 2e_3\Delta_2\varphi - \frac{\partial(ce_3r\varphi)}{\partial r} + 4\,e_3\varphi - \\ -\frac{\partial}{\partial r}[(ce_1-e_2)r^2\varphi] = \frac{\partial}{\partial r}[(k_{10}-ck_{20})T], \qquad (33)$$

$$c)\ c\frac{\partial^2\Psi_2}{\partial r^2} + \frac{2}{r}\frac{\partial\Psi_2}{\partial r} + \Delta_2\Psi_2 + \frac{ce_2-æ}{æ}\frac{\partial(r\Psi_1)}{\partial r} + \frac{2}{æ}\Psi_1 + \frac{1}{æ}\frac{\partial(e_3\varphi)}{\partial r} + \\ \frac{c_2+æ}{c_2æ}\frac{2}{r}(e_3\varphi) + e_1\frac{\partial\varphi}{\partial r} == \frac{\partial}{\partial r}k_{20}T,$$

$$d)\ \left(?_1 - \frac{e_3^2}{æ}\right)\Delta_2\varphi + \frac{\partial}{\partial r}\left\{r^2\left[?_2\frac{\partial\varphi}{\partial r} - \frac{e_2}{æ}\frac{\partial(e_3\varphi)}{\partial r} - \frac{2e_1}{æ}e_3\varphi\right]\right\} = \frac{\partial}{\partial r}(e_1\Delta_2\Psi_2) + \\ 2e_3\Delta_2\frac{\partial\Psi_2}{\partial r} + +\frac{\partial}{\partial r}\left[r^2\left(e_2\frac{\partial^2\Psi_2}{\partial r^2} + 2e_1\frac{\partial\Psi_2}{\partial r}\right)\right] + \frac{e_3}{æ}\frac{\partial}{\partial r}\Delta_2\Psi_1 + \\ \frac{\partial}{\partial r}\left[r^2\left(\frac{e_2}{æ}\frac{\partial^2\Psi_1}{\partial r^2} + 2\frac{e_1}{æ}\frac{\partial\Psi_1}{\partial r}\right)\right],$$

then for the proposed class of boundary-value problems of thermal electric elasticity the general solution in the class of regular functions is presented in the form:

$$u = \frac{\partial\Psi_2}{\partial r} + \frac{r}{æ}\Psi_1 + \frac{e_3}{æ}\varphi, \\ v = \frac{1}{r\sin\beta}\frac{\partial\Psi_2}{\partial\alpha} + \frac{1}{æ}\frac{\partial\Psi_0}{\partial\beta}, \\ w = \frac{1}{r}\frac{\partial\Psi_2}{\partial\beta} - \frac{1}{æ\sin\beta}\frac{\partial\Psi_0}{\partial\alpha}. \qquad (34)$$

Proof. If we consider (31) then (10b,c,d) can be rewritten in the following form:

$$\frac{\partial}{\partial\alpha}\left[K + \frac{2}{r}(e_3\varphi)\right] + \frac{\partial}{\partial r}\left[rB_\beta\sin\beta + \frac{\partial(e_3\varphi)}{\partial\alpha} - \sin\beta\frac{\partial^2(r\Psi_0)}{\partial r\,\partial\beta}\right] = 0, \\ \frac{\partial}{\partial\beta}\left[K + \frac{2}{r}(e_3\varphi)\right] - \frac{\partial}{\partial r}\left[rB_\alpha - \frac{\partial(e_3\varphi)}{\partial\beta} - \frac{1}{\sin\beta}\frac{\partial^2(r\Psi_0)}{\partial r\,\partial\beta}\right] = 0, \qquad (35)$$

$$\frac{\partial}{\partial r}\left[rB_\beta\sin\beta + \frac{\partial(e_3\varphi)}{\partial\alpha} - \sin\beta\frac{\partial^2(r\Psi_0)}{\partial r\,\partial\beta}\right] + \frac{\partial}{\partial r}\left[rB_\alpha - \frac{\partial(e_3\varphi)}{\partial\beta} - \frac{1}{\sin\beta}\frac{\partial^2(r\Psi_0)}{\partial r\,\partial\beta}\right] = 0.$$

From (35) follows that there exists function Ψ_1 such that:

$$a)\ K = \frac{\partial(r\Psi_1)}{\partial r} - \frac{2}{r}(e_3\varphi),$$

$$b)\ B_\alpha = \frac{\partial\Psi_1}{\partial\beta} + \frac{1}{r\sin\beta}\frac{\partial^2(r\Psi_0)}{\partial r\partial\alpha} + \frac{e_3}{r}\frac{\partial\varphi}{\partial\beta}, \qquad (36)$$

$$c)\ B_\beta = -\frac{1}{r\sin\beta}\frac{\partial\Psi_1}{\partial\alpha} + \frac{1}{r}\frac{\partial^2(r\Psi_0)}{\partial r\partial\beta} + \frac{e_3}{r}\frac{\partial\varphi}{\partial\alpha}$$

If (36) is introduced in (10a) we receive:

$$\frac{\partial}{\partial r}\left(cr^2\frac{\partial(r\Psi_1)}{\partial r}\right)-2r\frac{\partial(r\Psi_1)}{\partial r}+r^3\Delta_2\Psi_1+2e_3\Delta_2\varphi-\frac{\partial(ce_3r\varphi)}{\partial r}+4e_3\varphi-$$
$$-\frac{\partial}{\partial r}[(ce_1-e_2)r^2\varphi]=\frac{\partial}{\partial r}[(k_{10}-ck_{20})T]. \quad (37)$$

Using (36) and (31), from (11) we get:

$$\begin{aligned}
&\frac{\partial}{\partial\beta}\left(\mathrm{u}-\frac{r}{æ}\Psi_1-\frac{e_3}{æ}\varphi\right)-\frac{\partial}{\partial r}\left(r\,\mathrm{w}+\frac{r}{æ\sin\beta}\frac{\partial\Psi_0}{\partial\alpha}\right)=0,\\
&\frac{\partial}{\partial\alpha}\left(u-\frac{r}{æ}\Psi_1-\frac{e_3}{æ}\varphi\right)-\frac{\partial}{\partial\beta}\left(rv\sin\beta-\frac{r\sin\beta}{æ}\frac{\partial\Psi_0}{\partial\beta}\right)=0,\\
&\frac{\partial}{\partial\alpha}\left(r\,\mathrm{w}+\frac{r}{æ\sin\beta}\frac{\partial\Psi_0}{\partial\alpha}\right)-\frac{\partial}{\partial\beta}\left(rv\sin\beta-\frac{r\sin\beta}{æ}\frac{\partial\Psi_0}{\partial\beta}\right)=0.
\end{aligned} \quad (38)$$

From (38) follows that there is function such that:

$$\begin{aligned}
\mathrm{u}&=\frac{\partial\Psi_2}{\partial r}+\frac{r}{æ}\Psi_1+\frac{e_3}{æ}\varphi\\
\mathrm{v}&=\frac{1}{r\sin\beta}\frac{\partial\Psi_2}{\partial\alpha}+\frac{1}{æ}\frac{\partial\Psi_0}{\partial\beta}\\
\mathrm{w}&=-\frac{1}{r}\frac{\partial\Psi_2}{\partial\beta}+\frac{1}{æ\sin\beta}\frac{\partial\Psi_0}{\partial\alpha}
\end{aligned} \quad (39)$$

Substituting (39) and (36a) into (11) and (2), taking into account formulas (7), (8) we get:

$$c\frac{\partial^2\Psi_2}{\partial r^2}+\frac{2}{r}\frac{\partial\Psi_2}{\partial r}+\Delta_2\Psi_2+\frac{cc_2-æ}{æ}\frac{\partial(r\,\Psi_1)}{\partial r}+\frac{2}{æ}\Psi_1+\frac{1}{æ}\frac{\partial(e_3\varphi)}{\partial r}+\frac{c_2+æ}{æ}\frac{2}{r}(e_3\varphi)+e_1\frac{\partial\varphi}{\partial r}=k_{20}T,$$

Thus the theorem is proved as (39) coincides with (34) and (37), (39), (41) with (33b,c,d).

In the end one observation should be made. General temperature field can be presented as the superposition of two fields: solenoidal field when divU=0, u=0, T=0, φ=0 and thermo-electric elastic field with spherical rotor of shift vector when rot_rU=0.

References

[1] Nowacki W. *Electromagnetic effect in solid bodies*. Moskow, “Mir”, 1986.

[2] Grinchenko W.T., Ulitko A.F., Shulga N.A. *Electroelasticity*. Kiev, “Naukova Dumka” 1989.

[3] Khomasuridze N. Thermoelastic equilibrium of multilayer spherical bodies. *Seminar of I. Vekua Institute of Applied Mathematics*, 1997, Tbilisi, V. 23.

[4] Beitmen G., Erdei A., *The Higher transcendental Functions*, тт. 1,2, Moskow, “Nauka”, 1973.

[5] Tikhonov A.N., Samarski A.A., *Equations of Mathematical Physics*, Moskow, “Nauka”, 1986.

In: Several Problems of Applied Mathematics and Mechanics ISBN: 978-1-62081-603-5
Editors: Ivane Gorgidze and Tamar Lominadze

Chapter 19

A ONE-WAY FUNCTION AND KEY EXCHANGE ALGORITHM – A NEW APPROACH TO ASYMMETRIC CRYPTOGRAPHY

R. Megrelishvili
Iv. Javakhishvili Tbilisi State University, Tbilisi, Georgia

Abstract

In this paper, the author presents the original approach to synthesize the one-way function and the key exchange algorithm via an open channel. In this work, the reader will find the one-way function's structure, first published in [1]. Furthermore, this one-way function was studied and reviewed in several publications (see [2 - 4]). These studies show that it is possible to build large sets of n x n-sized commutative matrices over the Galois field GF(p) for fixed natural values of n. The set is built as a cyclic multiplicative group of cardinality $2^n - 1$ with a specific generator-matrix. These sets of matrices are used in the construction of the one-way function as well as in the key exchange algorithm. Section 2 of the article represents the principally new one-way function, and section 3 addresses the new key exchange algorithm via an open channel.

Keywords: Public channel, one-way function, key exchange, Galois field $\mathrm{GF}\left(2^n\right)$

1. Introduction

The Diffie-Hellman algorithm for a key exchange via the open communication channel (analong with the RSA algorithm) created a new direction of cryptography, namely, asymmetric cryptography [5, 6]. With the advent of a new direction the importance and level of research in cryptography has decidedly increased. Number theory and structures of algebraic theory are the basis for this fundamental change in cryptography.

There is a well-known exponential function in number theory (i.e., the one-way function: $y = g^x$ mod p). It is essential that, if p is a prime number of high order, for example, $p>2^{100}$, while the other parameters are defined with the condition that $1 < g, x, y < p$, then it is easily seen that y is easily calculated, if values of thc x, p and g are fixed. Nonetheless, it is impossible to calculate the value of x in real time, if the parameters p, g and y are fixed, since this calculation is associated with an exhaustive possibility of options, that is, it can't be performed in real time. Please note that this is correct if the parameter p is a prime number of high order.

This fact is known as the Discrete Logarithm Problem, and it is based on the above-mentioned one-way function: $y = g^{x} \bmod p$. The latter is a one-way function which defines the Diffie-Hellman algorithm [5]. In reality, the function $y = g^x$ (without module p) is a usual reversible function (direct function y=f(x) and reverse function x=t(y)) and the functions $y = g^x$ and $x = \log_g y$ can be calculated without problems. However, if the latter is treated as a comparison mod p, then the calculation of function $x = \log_g y$, for a given $y = g^x \bmod p$, is impossible in a real time period. In such a case, this function is treated as a one-way function ($y = g^x \bmod p$ is a one-way function) at its best qualities.

A new structure for the one-way function is described in this paper and, accordingly, an algorithm for a key exchange via an open channel is also addressed. This structure of a one-way function is fundamentally different from the above-described structure of a one-way function. These issues are discussed in sections 2 and 3.

2. A New One-way Function

The Diffie-Hellman Algorithm (as was explained in the introduction), based on the exponential one-way function, allowed the authors to perform a key exchange without a secret channel. This result comes from special properties of the mentioned one-way function. Because of this, asymmetric methods have a great advantage compared to the symmetric methods. Despite this, there is some equilibrium between asymmetric and symmetric methods if we consider the situation in practice.

What is the reason? Mainly, symmetric methods are a thousand times faster than asymmetric methods [7]. This is the case because the one-way functions (i.e., the Diffie-Hellman and Rivest-Shamir-Adleman algorithms) are exponential functions and, therefore, these operations (corresponding to exponential functions) require much more time than other operations (such as addition and multiplication).

Before we consider our one-way function, we say that our one-way function, based on matrix operations, contains only the operations of addition and multiplication; therefore, it is much faster than the function with which we will compare our new one-way function.

Unlike according to the Diffie-Hellman protocol, as an alternative approach to constructing one-way functions, the author uses n x n matrices over Galois fields GF(p). The idea of this alternative approach is as follows:

Let A be the set of commutative matrices (suppose that the cardinality of A is approximately 10^{30} [i.e., card (A) = 2^{100}], which equals the lowest steadiness level of contemporary cryptosystems).

Also, let the matrices

$$A^{2^0}, A^{2^1}, \ldots, A^{2^{n-1}} \tag{1}$$

create the basis of the set A, where $A^{2^i} \neq A^{2^j}$, if $i \neq j$ and A is the generator-matrix. Therefore, any matrix A_i of the set A is obtained as the linear combination of matrices (1):

$$Ai = c_0 A^{2^0} + c_1 A^{2^1} + \ldots + c_{n-1} A^{2^{n-1}}, \tag{2}$$

where $c_i \in GF(2)$.

In order to reach the minimal level of contemporary cryptographic steadiness, we should consider only those sets of matrices which have cardinalities above 2^{100}. For example, if matrices are given over the GF(2) field, then their size must be greater than 100 x 100. We're talking about a set of matrices, so as to draw an analogy between the set of matrices A and the set of natural numbers, given with the prime number p of the Diffie-Hellman algorithm.

The one-way function's sustainability of the Diffie-Hellman algorithm depends on the high order p; therefore, the high cardinality of A provides sustainability of our one-way function.

Our one-way function is as follows:

$$\nu A = u, \tag{3}$$

where $\nu, u \in V_n$ (V_n is the vector space over GF(2)), A is the set of matrices of (2). The substantive issues of our one-way function's operation are discussed in Section 3.

3. A New One-way Function and Key Exchange Algorithm

Now, we'll formulate our key exchange scheme:

Let the basis (1) or the generator-matrix A be public as well as the vector $\nu = (\upsilon_1, \ldots, \upsilon_n)(\upsilon_i \in GF(2))$.

Imagine two people, Alice and Bob, forming a secret exchange key via the public channel:

- Alice chooses the n x n-sized secret matrix $A_1 \in A$ and sends the following vector to Bob:

$$u_1 = \nu A_1; \tag{4}$$

- Bob chooses the n x n-sized secret matrix $A_2 \in A$ and sends Alice the vector:

$$u_2 = \nu A_2; \tag{5}$$

- Alice computes $k_1 = u_2 A_1$;
- Bob computes $k_2 = u_1 A_2$.

The crucial point is that k_1 and k_2 are the same secret exchange keys, as $k_1 = \mathrm{v}(A_2 A_1) = \mathrm{v}(A_1 A_2) = k_2$ (here, the author uses set A's commutability).

Consider a simple example of the algorithm's functioning. Suppose, A is an n x n dimensional set of matrices, where n=3:

$$A = \begin{pmatrix} 1 & 0 & 1 \\ 1 & 0 & 0 \\ 0 & 1 & 1 \end{pmatrix},\ A^2 = \begin{pmatrix} 1 & 1 & 0 \\ 1 & 0 & 1 \\ 1 & 1 & 1 \end{pmatrix},\ A^3 = \begin{pmatrix} 0 & 0 & 1 \\ 1 & 1 & 0 \\ 0 & 1 & 0 \end{pmatrix}, \ldots,\ A^7 = \begin{pmatrix} 1 & 0 & 0 \\ 0 & 1 & 0 \\ 0 & 0 & 1 \end{pmatrix} \tag{6}$$

Suppose that Alice selects matrix A^2 as the secret matrix (see (6)); Bob, for his part, chooses the matrix A^3; we also assume that v=(110). Then, therefore, our algorithm will:

- Alice computes and sends to Bob the following vector:

$$u_1 = vA_1 = (011);$$

- Bob computes and sends to Alice the following vector:

$$u_2 = vA_2 = (111);$$

- Alice computes the exchanged key $k_1 = u_2 A_1 = (100)$;
- Bob computes the exchanged key $k_2 = u_1 A_2 = (100)$.

As we can see, $k_1 = k_2$ and the results are correct.

Although the results are correct, we are particularly interested in the notion of the matrices' recursive dependency (this is the relationship between the rows or columns of the matrix) [2-4]. We attached particular importance to dependences of this type because they determine the suitability or unsuitability of the class of A matrices for our key exchange algorithm for an open channel of communication.

This recursive relationship of matrices is important to our one-way function as the module p (mod p) for the one-way function of the Diffie-Hellman algorithm [5] (this issue, i.e., the one-way function of Diffie-Hellman, is discussed in the first section of this paper).

We can conclude the following: our function (3) may be useful (and with more efficiency) as a one-way function to a special class of matrices, free from recursive dependency; nonetheless, the exact same function (3) can't be used as a one-way function for the class of matrices with recursive dependency (there is a certain analogy: the Diffie-

Hellman function $g^x = y$ mod p can be useful as a one-way function, but the exact same function $g^x = y$ without the module p can't be used as a one-way function).

Thus, it is possible to break this algorithm, but what constitutes a recursive dependence of matrices? This is question associated with analyzing algebraic structures. Some authors believe that this is only due to the concept of companion matrices but, from the standpoint of the author of this paper, this is not true [2-4].

References

[1] R. Megrelishvili, M. Chelidze, K. Chelidze, *On the construction of secret and public key cryptosystems*, Iv. Javakhishvili Tbilisi State University, I. Vekua Institute of Applied Mathematics, Informatics and Mechanics (AMIM), v. 11, No 2, 2006, pp. 29-36.

[2] R. Megrelishvili, A. Sikharulidze, New matrix sets generation and the cryptosystems, *Proceedings of the European Computing Conference and the Third International Conference on Computational Intelligence*, Tbilisi, Georgia, June 26-28, 2009, pp. 253-255.

[3] R. Megrelishvili, M. Chelidze, G. Besiashvili, Investigation of new matrix-key function for the public cryptosystems, *The Third International Conference "Problems of Cybernetics and Information"*, v. 1, September 6-8, Baku, Azerbaijan, 2010, pp. 75-78.

[4] R. Megrelishvili, M. Chelidze, G. Besiashvili, One-way matrix function – analogy of Diffie-Hellman protocol, *Proceedings of the Seventh International Conference*, IES – 2010, 28 September – 3 October, Vinnytsia, Ukraine, 2010, pp. 341- 344.

[5] W. Diffie and M.E. Hellman, New Direction in Cryptography, *IEEE Transaction on Information Theory*, IT-22, n. 6, Now. 1996, pp. 644-654.

[6] R.L. Rivest, A. Shamir, L.M. Adleman, *A Method for Obtaining Digital Signatures and Public-key Cryptosystems, Communications of the ACM*, v. 21, n. 2, Feb. 1978, pp. 120-126.

[7] B. Schneier, *Applied Cryptography*, John Wiley and Sons, Inc. New York, 1996.

INDEX

A

B

C

D

E

P

R

S

T

U

V

W